Peptides For Beginners

Achieve Optimal Health with Safe Protocols and Simple Dosage Guide for Muscle Growth, Fat Loss, Longevity and Anti-Aging, Immunity, and Cognitive Performance

Earl Fischer

DISCLAIMER

The content provided in this book is intended for informational and educational purposes only. It is not meant to substitute professional medical advice, diagnosis, or treatment. The author and publisher do not claim to offer medical, legal, or professional advice, and readers are advised to consult with a qualified healthcare provider before making any decisions or taking actions based on the information contained in this book.

While every effort has been made to ensure the accuracy and reliability of the content, the author and publisher make no guarantees about the completeness, up-to-date nature, or potential errors in the material. The information presented is based on research and personal experiences but is not reviewed or endorsed by medical authorities such as the FDA or any equivalent medical agency/ authority.

The use of any peptides, protocols, or recommendations discussed in this book should be undertaken at the reader's discretion and risk. It is strongly advised that individuals consult a healthcare professional, especially those who are pregnant, nursing, taking medications, or managing chronic health conditions. Results may vary, and the information provided should not be viewed as a guarantee or prescription.

The opinions expressed in this book are solely those of the author and may not reflect the views of any organizations or institutions. The author and publisher disclaim any liability for any loss, injury, or damage incurred as a result of applying the information provided herein.

By reading this book, you acknowledge and agree that the author and publisher are not responsible for any outcomes resulting from the application of this material. Unauthorized reproduction, distribution, or transmission of this content, in any form, is prohibited without prior written consent from the author.

Table of Contents

INTRODUCTION 11
CHAPTER 1. INTRODUCTION TO PEPTIDES 12
   1.1 What Are Peptides? 12
   1.2 History and Evolution of Peptides in Medicine 12
   1.3 Difference Between Peptides and Proteins 12
   1.4 Natural vs. Synthetic Peptides 13
   1.5 Peptide Synthesis Techniques 14
CHAPTER 2. THE SCIENCE BEHIND PEPTIDES 15
   2.1 Structure and Function of Peptides 15
   2.2 How Peptides Work in the Body 15
   2.3 Types of Peptides 15
      2.3.1 Oligopeptides 15
      2.3.2 Polypeptides 16
      2.3.3 Cyclic Peptides 16
   2.4 Key Peptide Receptors and Pathways 16
   2.5 The Role of Amino Acids in Peptide Functionality 16
CHAPTER 3. HOW TO START USING PEPTIDES 17
   3.1 Choosing the Right Peptide for Your Needs 17
   3.2 How to Purchase Peptides Safely 17
   3.3 How to Administer Peptides 18
      3.3.1 Injections 18
      3.3.1.1 Step-by-Step Guide to Reconstitute CJC-1295 for Injection 18
      3.3.2 Oral Capsules 20
      3.3.3 Nasal Sprays 20
   3.4 Dosage Guidelines and Cycling Peptides 20
   3.5 Common Challenges and How to Overcome Them 21
   3.6 Common Mistakes to Avoid When Starting Peptides 22
CHAPTER 4. SAFETY AND REGULATIONS 23
   4.1 Peptide Safety: Understanding Side Effects and Risks 23

4.2 Legal and Regulatory Considerations in Peptide Usage ... 24

4.3 Peptides and FDA: Current Status of Approval ... 25

CHAPTER 5. THERAPEUTIC PEPTIDES AND USES ... 26

5.1 Peptides for Fat Loss ... 26

    Ipamorelin ... 26

    AOD-9604 ... 26

    Semaglutide ... 27

    Tirzepatide ... 28

    Tesofensine ... 28

    Tesamorelin ... 29

    MOTS-C ... 29

    5-Amino 1MQ ... 30

5.2 Peptides for Muscle Growth and Performance ... 30

    Sermorelin ... 31

    BPC-157 ... 31

    TB-500 ... 32

    IGF-1 LR3 ... 33

    DSIP ... 34

    GHRP-2 ... 35

    GHRP-6 ... 36

    Hexarelin ... 37

    PEG-MGF ... 38

    MK-677 ... 38

    Ipamorelin ... 39

    CJC-1295 ... 40

5.3 Peptides for Brain Health and Cognitive Performance ... 41

    Semax ... 41

    Selank ... 42

    Dihexa ... 43

    Cerebrolysin ... 43

    Orexin A ... 44

    PE-22-28 45

    FGL 46

5.4 Peptides for Longevity and Anti-Aging 47

    Epitalon 47

    Thymalin 48

    GHK-Cu 49

    Humanin 50

    TB-4/TB-500 51

5.5 Peptides for Sexual Health 52

    PT-141 52

    Kisspeptin 53

    Melanotan II 54

5.6 Peptides for Immunity 55

    Thymosin Alpha-1 55

    LL-37 56

    VIP 57

    KPV 58

    ARA-290 58

    SS-31 59

5.7 Peptides for Sleep 60

    DSIP (Delta Sleep-Inducing Peptide) 60

    Epitalon 61

    Thymosin Beta-4 62

5.8 Peptides for Skin, Hair, and Aesthetics 63

    GHK-Cu 63

    Argireline 64

    PTD-DBM 64

    BPC-157 65

    Melanotan I & II 66

5.9 Peptides for Women 66

    Kisspeptin 66

- Peptides for Menopause 67
  - PT-141 68
- 5.10 Peptides for Men 68
  - Gonadorelin 69
  - Kisspeptin 69
  - PT-141 70

CHAPTER 6. PEPTIDE STACKS AND COMBINATIONS 71

- 6.1 Peptides Stacks/Combos for Fat Loss 71
  - Ipamorelin + CJC-1295 71
  - Ipamorelin + CJC-1295 + AOD-9604 71
  - Semaglutide + MOTS-C + Tesamorelin 72
  - Tirzepatide + Tesofensine + 5-Amino 1MQ 73
  - Tesamorelin + CJC-1295 + MK-677 73
  - AOD-9604 + Ipamorelin + Tirzepatide 74
- 6.2 Peptides Stacks/Combos for Muscle Growth 74
  - CJC-1295 + Ipamorelin + IGF-1 LR3 74
  - CJC-1295 + Ipamorelin + BPC-157 75
  - CJC-1295 + GHRP-2 + BPC-157 75
  - CJC-1295 + GHRP-6 + BPC-157 76
  - MK-677 + GHRP-6 + PEG-MGF 77
  - TB-500 + BPC-157 + CJC-1295 77
  - IGF-1 DES + Follistatin-344 + GHRP-2 78
  - Hexarelin + Ipamorelin + IGF-1 LR3 78
  - Hexarelin + TB-500 + PEG-MGF 79
- 6.3 Brain Health and Cognitive Performance Stacks/Combos 79
  - Semax + Selank + Cerebrolysin 79
  - Semax + Selank + Dihexa 80
  - Dihexa + Selank + FGL 80
  - Cerebrolysin + Semax + Epitalon 81
  - Epitalon + Selank + Dihexa 82
  - Semax + CJC-1295 + GHRP-2 82

Dihexa + Orexin A + FGL 83

Semax + PE-22-28 + Orexin A 83

6.4 Peptides Stacks/Combos for Longevity and Anti-Aging 84

Epitalon + Thymalin + GHK-Cu 84

Epitalon + BPC-157 + TB-500 85

Epitalon + Humanin + GHK-Cu 85

MOTS-C + Humanin + SS-31 (Elamipretide) 86

Epitalon + CJC-1295 + GHRP-2 87

GHK-Cu + BPC-157 + TB-500 87

Thymalin + Epitalon + GHRP-6 88

6.5 Peptides Stacks/Combos for Sexual Health 88

PT-141 + Kisspeptin + Melanotan II 88

PT-141 + CJC-1295 + Ipamorelin 89

Gonadorelin + PT-141 + MK-677 90

Kisspeptin + CJC-1295 + Ipamorelin 90

PT-141 + Melanotan II + CJC-1295 91

6.6 Peptides Stacks/Combos for Immunity 91

Thymosin Alpha-1 + LL-37 + VIP 91

Thymosin Alpha-1 + BPC-157 + SS-31 92

VIP + LL-37 + SS-31 92

Thymosin Alpha-1 + KPV + ARA-290 93

Thymosin Alpha-1 + LL-37 + BPC-157 94

6.7 Peptides Stacks/Combos for Skin, Hair and Aesthetics 94

GHK-Cu + BPC-157 + Epitalon 94

GHK-Cu + PTD-DBM + Argireline 95

GHK-Cu + CJC-1295 + Ipamorelin 96

BPC-157 + GHRP-2 + GHK-Cu 96

6.8 Key Considerations for Peptide Combos/Stacking 97

CHAPTER 7. PEPTIDES AND LIFESTYLE 98

7.1 Nutrition, Exercise and Recovery 98

7.1.1 Nutrition 98

7.1.2 Exercise 98

7.1.3 Recovery 99

7.2 Managing Your Expectations 99

7.2.1 Short-Term Benefits (Within Days to Weeks) 99

7.2.2 Long-Term Benefits (Within Months) 100

7.2.3 Balancing Expectations 100

CHAPTER 8. CONCLUSION 101

8.1 Resources for Further Learning and Research 101

References 103

INTRODUCTION

Peptides are quickly becoming popular in the field of regenerative medicine due to their ability to promote healing and repair tissues at a cellular level. Unlike many traditional treatments, which tend to mask symptoms, peptides work by addressing the root causes of damage or degeneration, allowing the body to heal itself more effectively.

Found naturally in the human body and also synthesized for specific purposes, peptides are used to stimulate cell migration, promote recovery and tissue regeneration. These regenerative peptides have gained popularity among athletes and fitness enthusiasts because they help speed up recovery from sports injuries and intense training.

However, their benefits extend beyond athletes, as they are also used in treating conditions like chronic pain, arthritis, hormonal imbalances, erectile dysfunction and inflammatory diseases. With more research, peptides are likely to become even more integral in developing treatments for age-related degeneration, allowing individuals to recover from injuries faster and experience less wear and tear as they age. Chronic diseases like diabetes, heart disease, and neurodegenerative conditions are some of the most pressing health issues worldwide. Peptides offer new possibilities in the treatment and management of these conditions. Peptides also play a significant role in slowing down the effects and even reversing certain aspects of cellular aging.

This book serves as a beginner-friendly guide to understanding peptides, their uses, and how they can benefit your health. While peptides may sound complex, their applications are straightforward and easy to incorporate into everyday life. You will learn what peptides are, how they work in the body, and how they are being applied in modern healthcare. Each peptide has unique properties, and the right choice depends on your individual health needs and goals.

Safety is a key focus throughout this book. While peptides are generally considered safe when used correctly, they must be handled and administered with care. This book includes practical advice on how to source and prepare peptides, administer them, and monitor their effects. It also provides information on potential risks and side effects, helping you to make informed decisions.

CHAPTER 1. INTRODUCTION TO PEPTIDES

1.1 What Are Peptides?

Peptides are short chains of amino acids. Think of them as tiny building blocks that make up the proteins in your body. While proteins are long, complex chains of these amino acids, peptides are much smaller and simpler. They usually consist of between 2 and 50 amino acids linked together in a specific sequence.

Your body naturally produces many different peptides, and they play essential roles in various biological processes. Peptides can act as signals between cells, helping to regulate activities such as healing, growth, and metabolism. They can also serve as hormones, transporting information between organs and tissues.

In recent years, peptides have gained a lot of attention in medicine, fitness, and wellness communities. This is because scientists have found ways to create synthetic peptides that can mimic the body's natural peptides. These synthetic versions can be used to treat various health conditions, enhance physical performance, or even slow down the effects of aging.

1.2 History and Evolution of Peptides in Medicine

The use of peptides in medicine is not a brand-new concept. In fact, peptides have been studied and used for nearly a century. The first known medical peptide was insulin, which was discovered in the early 1920s. Insulin, a peptide hormone, revolutionized the treatment of diabetes, allowing millions of people worldwide to manage their blood sugar levels effectively.

Since then, researchers have developed a wide range of therapeutic peptides. In the early years, most of the focus was on naturally occurring peptides, but as technology advanced, scientists began creating synthetic versions. These synthetic peptides often work more efficiently or target specific functions within the body. For example, synthetic peptides like BPC-157 or TB-500 are popular in the world of sports and rehabilitation for their ability to accelerate healing.

In the 21st century, peptides have moved from being a niche therapy to something that's becoming more mainstream. With over 800 peptide drugs currently in development and many already available on the market, peptides are expected to play a major role in the future of healthcare.

1.3 Difference Between Peptides and Proteins

Peptides and proteins both consist of amino acids, but the main difference between them is their size. Peptides are shorter, typically made of up to 50 amino acids, while proteins are much larger and can contain thousands of amino acids.

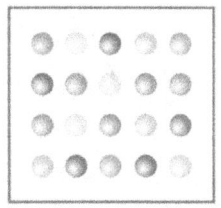 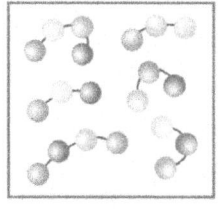 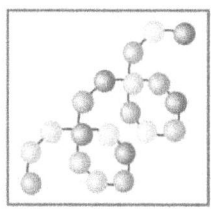

Amino acids — Peptides — Protein

Another key difference is how they function. While peptides often act as signaling molecules or hormones, proteins tend to serve more structural roles in the body. For instance, collagen, which gives your skin and tissues their strength, is a protein. On the other hand, insulin, which helps regulate blood sugar levels, is a peptide hormone.

Additionally, peptides tend to be more versatile in medical applications. They are smaller and easier to manipulate in labs, which makes them easier to study and use in treatments. This is why there is such growing interest in developing peptide-based therapies for everything from weight loss to cognitive enhancement.

1.4 Natural vs. Synthetic Peptides

Peptides can be found naturally in the body, or they can be made in a lab. Natural peptides are produced by your cells and help regulate a variety of functions, such as repairing damaged tissues, regulating hormones, and controlling metabolism.

Natural Peptides

These are the peptides your body produces on its own. Every day, your cells make thousands of different peptides that keep your body running smoothly. Some examples include:

- **Insulin:** Regulates blood sugar levels.
- **Oxytocin:** Plays a role in childbirth and bonding between people.
- **Glucagon:** Helps raise blood sugar levels when they're too low.

Synthetic Peptides

Scientists create synthetic peptides in the lab. These peptides are designed to mimic the natural peptides in your body or improve upon them in some way. For example, synthetic peptides like CJC-1295 and Ipamorelin are used to stimulate the body's production of growth hormone, helping people build muscle, recover faster, and even slow down aging.

Because synthetic peptides are made in a controlled environment, researchers can tweak them for specific uses. This opens up a lot of possibilities for treating different health issues or enhancing performance in a way that natural peptides might not be able to do on their own.

1.5 Peptide Synthesis Techniques

To make peptides in the lab, scientists use a process called **peptide synthesis**. There are two main methods used to create synthetic peptides: **solid-phase peptide synthesis (SPPS)** and **liquid-phase peptide synthesis (LPPS)**.

- **Solid-Phase Peptide Synthesis (SPPS):** This is the most common technique for creating peptides. In SPPS, the peptide chain is built one amino acid at a time while attached to a solid surface. This method is preferred because it's efficient and allows scientists to create peptides of various lengths and complexities.

- **Liquid-Phase Peptide Synthesis (LPPS):** LPPS is used less frequently but can be better for making longer, more complicated peptides. The process occurs in a solution rather than on a solid surface. It's more time-consuming, but in certain cases, it produces better results.

Both methods involve linking amino acids together in a specific sequence to create the desired peptide. Once the peptide is complete, it is purified and tested to make sure it functions as expected.

CHAPTER 2. THE SCIENCE BEHIND PEPTIDES

This is not a science class; however, I will try to explain the science behind these ever-powerful peptides. It is fascinating to know how peptides work in our bodies.

2.1 Structure and Function of Peptides

2.1.1 Structure

Peptides are composed of amino acids linked together in a specific sequence, forming short chains. These chains fold into three-dimensional shapes that determine their function in the body. The sequence of amino acids dictates how the peptide interacts with other molecules and receptors. They range from just a few amino acids (like dipeptides or tripeptides) to around 50 amino acids. The specific arrangement and folding of these amino acids give each peptide its unique properties and functions.

2.1.2 Functionality

- **Signaling:** Peptides act as messengers between cells, transmitting signals that regulate biological processes such as growth, metabolism, and immune response.
- **Hormones:** Many peptides function as hormones, controlling activities like insulin regulation (important for blood sugar management) and growth hormone release (crucial for muscle growth and repair).
- **Enzymes:** Some peptides act as enzymes, speeding up chemical reactions in the body that are necessary for digestion, metabolism, and other vital processes.

2.2 How Peptides Work in the Body

Peptides exert their effects by binding to specific receptors on cell surfaces or inside cells. This binding triggers a cascade of biochemical reactions that regulate various biological processes. For example:

- **Cell Communication:** Peptides can relay messages between cells, instructing them to perform specific actions like releasing hormones or activating immune responses.
- **Receptor Activation:** By binding to receptors, peptides can initiate or inhibit physiological responses such as muscle contraction, inflammation, or neurotransmitter release.
- **Enzymatic Activity:** Peptides can act as catalysts, increasing the rate of chemical reactions that break down molecules or build new ones essential for cellular function.

2.3 Types of Peptides

2.3.1 Oligopeptides

These are short chains of amino acids, typically consisting of 2 to 20 amino acids. Oligopeptides include dipeptides (2 amino acids) and tripeptides (3 amino acids), and they often act as signaling molecules or precursors to larger peptides and proteins.

2.3.2 Polypeptides

Polypeptides are longer chains of amino acids, ranging from 20 to 50 amino acids in length. They are more complex than oligopeptides and can have diverse functions, including hormone regulation, enzymatic activity, and structural support in tissues.

2.3.3 Cyclic Peptides

Cyclic peptides have a unique structure where the amino acid chain forms a closed loop. This cyclic structure enhances their stability and resistance to degradation, making them valuable in drug development and therapeutic applications.

2.4 Key Peptide Receptors and Pathways

Peptides exert their effects by binding to specific receptors on cell surfaces or inside cells. These receptors are proteins that recognize and respond to the presence of peptides, initiating cellular processes or signaling cascades.

G-protein Coupled Receptors (GPCRs):

Many peptides bind to GPCRs, a large family of receptors involved in diverse physiological functions such as neurotransmission, hormone regulation, and sensory perception. GPCRs play an important role in mediating the effects of peptides on cellular activities.

Tyrosine Kinase Receptors:

Some peptides interact with tyrosine kinase receptors, which are involved in cell growth, differentiation, and metabolism. Binding of peptides to these receptors can activate signaling pathways that regulate cellular processes like growth and repair.

2.5 The Role of Amino Acids in Peptide Functionality

Amino acids are the building blocks of peptides and proteins, and their sequence determines the structure and function of peptides. Different amino acids contribute unique properties to peptides, influencing their stability, binding affinity, and biological activity.

Essential vs. Non-essential Amino Acids:

Essential amino acids cannot be synthesized by the body and must be obtained through diet. They play critical roles in peptide structure and function. **Non-essential** amino acids can be synthesized by the body and also contribute to peptide stability and function.

CHAPTER 3. HOW TO START USING PEPTIDES

3.1 Choosing the Right Peptide for Your Needs

When considering peptide therapy, the first step is to identify the specific health goals or issues you want to address. Since peptides target a wide range of functions ranging from fat loss and muscle growth to cognitive enhancement and sexual health. It is important to match the right peptide to your needs. Choosing the wrong peptide might not yield the desired results or could even lead to unwanted side effects.

To begin, think about the particular outcomes you are seeking. For example:

- **For muscle growth and recovery:** Peptides like **Ipamorelin** or **IGF-1 LR3** are good choices, as they boost growth hormone production and support tissue repair.
- **For fat loss: AOD-9604** or **Semaglutide** can help by enhancing fat metabolism and suppressing appetite.
- **For skin rejuvenation: GHK-Cu** is excellent for improving skin elasticity, reducing wrinkles, and speeding up wound healing.
- **For cognitive enhancement: Semax** or **Dihexa** might be your best options, as they support memory, focus, and overall brain health.

It is also important to consider any underlying health conditions or medications you are taking, as some peptides may interact with other treatments or affect specific conditions. Consulting with a healthcare professional who has experience with peptide therapy can be invaluable. They can help determine which peptide will work best for your individual needs and guide you through the process of starting your peptide regimen.

3.2 How to Purchase Peptides Safely

Purchasing peptides can be tricky since the market is largely unregulated, and there are many companies offering products of varying quality. To ensure that you are buying safe and effective peptides, it's important to do your research and choose a reputable supplier. Here are some key factors to consider:

- **Purity:** The most important aspect when purchasing peptides is their purity. High-purity peptides are more effective and safer. Look for suppliers that provide certificates of analysis (COAs) from independent third-party labs. These COAs will confirm the peptide's purity and ensure that the product is free of contaminants or harmful additives.
- **Reputation and Reviews:** Choose suppliers with a strong reputation in the industry. Read customer reviews, check online forums, and ask for recommendations from trusted sources who have experience with peptides. Reliable suppliers often have a solid track record and offer customer support to answer any questions you might have.

- **Transparent Labeling and Ingredient Lists:** Ensure that the supplier provides clear, accurate labeling on their products. Look for information about the peptide's concentration, dosage instructions, and expiration date. Avoid products that don't clearly disclose this information, as they may be counterfeit or low-quality.
- **Storage and Shipping:** Peptides are delicate compounds that require proper storage to maintain their potency. Most peptides should be stored in cool, dark environments (often refrigerated). Before purchasing, make sure the supplier follows proper shipping protocols, such as using insulated packaging or cold packs to prevent the peptides from degrading during transit.
- **Legal Considerations:** Depending on your country or region, the legal status of peptides may vary. Some peptides are available only with a prescription, while others can be purchased freely online. Make sure you understand the legalities of purchasing and using peptides in your area or field of work to avoid potential issues.

3.3 How to Administer Peptides

Once you've chosen the right peptide and purchased it from a reputable source, the next step is administering it correctly. Peptides can be administered in several ways, depending on the type of peptide and its intended use. The most common methods include injections, oral capsules, and nasal sprays.

3.3.1 Injections

The majority of peptides are administered via subcutaneous injection, meaning the peptide is injected just under the skin. This method ensures that the peptide enters the bloodstream quickly and begins working almost immediately. Administering injections can seem intimidating at first, but with proper technique, it is safe and relatively simple. Here's how to do it:

i. Use a sterile syringe and draw up the recommended dose of the peptide.
   Note: Clean the vial's rubber stopper with an alcohol swab before drawing up the solution to avoid contamination.
ii. Pinch a small area of skin, usually around the abdomen or thigh and clean with alcohol swab.
iii. Insert the needle at a 45-degree angle and slowly inject the peptide.
iv. Dispose the syringe safely in a sharps/needle container.

Injections are the most effective way to deliver peptides because they bypass the digestive system, which can break down peptides and reduce their effectiveness.

3.3.1.1 Step-by-Step Guide to Reconstitute CJC-1295 for Injection

1. Gather Supplies:

- **CJC-1295 Vial**
- **Bacteriostatic Water:** Used to mix with the peptide. This water contains a small amount of benzyl alcohol to keep it sterile after opening.

- **10 mL mixing syringe**
- **Insulin Syringe (1 ml)**: 30–100 unit syringes work best for dosing.
- **Alcohol Swabs**: To clean the vial tops and the injection area.

2. Prepare the CJC-1295 Vial and Bacteriostatic Water

- Take the **alcohol swab** and wipe the rubber stopper on top of the CJC-1295 vial to keep it sterile.
- Also, wipe the rubber stopper on the bacteriostatic water vial.

3. Draw Bacteriostatic Water into the Syringe

- Using your 10 mL mixing syringe, draw the desired amount of **bacteriostatic water** into the syringe. For a **5 mg** vial of CJC-1295, **5 ml of bacteriostatic water** is a common amount to use for reconstitution, as it makes measuring doses easier.

However, it is important to follow the instructions provided by the manufacturer of the peptide as they may have specific instructions for reconstitution.

4. Mix the Bacteriostatic Water with CJC-1295

- Insert the syringe into the **CJC-1295 vial** at a slight angle and slowly push the plunger to release the bacteriostatic water. Let the water run down the side of the vial to avoid direct contact with the powder, which can cause foaming or damage the peptide. Pull the syringe out.
- **Do not shake the vial.** Instead, gently swirl or roll the vial between your hands to help the powder dissolve. The peptide should mix smoothly with the water after a few minutes.

5. Calculate the Dosage for Injection

- After reconstituting with bacteriostatic water, your CJC-1295 solution will contain **1000 mcg per 0.1 ml (10 units)**.

So, to get a dose of **1000 mcg**, draw **10 units on the insulin syringe** to inject 1000 mcg.

6. Draw the Dose for Injection

- Wipe the rubber stopper on the reconstituted CJC-1295 vial with an alcohol swab.
- Flip the **CJC-1295 Vial** upside down, then insert the insulin syringe into the vial, and draw up **10 units (0.1 ml)** of the mixed solution to achieve the 1000 mcg dose.

7. Inject the Peptide (Subcutaneous Injection)

- Use an alcohol swab to clean the injection site, typically on the abdomen about 2 inches away from the belly button.
- Pinch a small section of skin, insert the needle at a 45-degree angle, and slowly inject the peptide.

3.3.2 Oral Capsules

Some peptides are available in oral form, but this is less common. Peptides are typically large molecules that are broken down by stomach acids before they can be absorbed into the bloodstream. However, advancements in peptide formulation have allowed certain peptides to be administered orally, such as **BPC-157** or **GLP-1 agonists** like **Semaglutide**. These capsules are convenient and easy to use but may be less effective than injections, as the body may not absorb them efficiently.

3.3.3 Nasal Sprays

Another method of peptide administration is through nasal sprays. Peptides like **Semax** or **Selank** are often delivered this way because the nasal cavity allows for fast absorption into the bloodstream without injections. Nasal sprays are user-friendly and non-invasive, making them a good option for people who are uncomfortable with needles. Simply spray the prescribed dose into one or both nostrils, and the peptide will be absorbed through the nasal tissues.

3.4 Dosage Guidelines and Cycling Peptides

Getting the dosage right is essential for the effectiveness and safety of peptide therapy. Overdosing can lead to unwanted side effects, while underdosing may result in minimal or no benefits. Since peptide dosage varies depending on the type of peptide, your health goals, and your individual body chemistry, follow recommended dosage guidelines or consult with a healthcare professional.

- **Start Low and Go Slow:** If you are new to peptides, it is a good idea to start with a low dose and gradually increase it. This allows your body to adjust and reduces the risk of side effects. For example, a typical starting dose for **Ipamorelin** might be around 100–200 mcg per injection, taken 1–2 times per day.

- **Timing:** The timing of peptide administration is also important. Some peptides, like those used for muscle recovery, are best taken post-workout, while others, like sleep-enhancing peptides, should be taken before bedtime. For peptides that stimulate growth hormone release, such as **CJC-1295** and **Ipamorelin**, it's often recommended to take them on an empty stomach, as food can interfere with their effectiveness.

- **Cycling Peptides:** To avoid building up a tolerance or desensitization to peptides, it's important to cycle them. This means using the peptide for a set period, such as 4–8 weeks, followed by a break. Cycling not only prevents your body from adapting to the peptide but also gives your system time to reset and maintain its natural balance. For example, with peptides like **GHK-Cu** or **BPC-157**, you may use them consistently for healing purposes and then take a break once the desired effect is achieved.

  Cycling is especially important with peptides that affect hormone levels, such as **growth hormone-releasing peptides**. Long-term, uninterrupted use of these peptides could lead to hormone imbalances or diminished results over time. Be mindful of the need to cycle peptides and take breaks as needed to maximize their benefits and minimize potential risks.

3.5 Common Challenges and How to Overcome Them

Starting and sticking with peptide therapy can come with a few challenges, especially for beginners. Here are some common challenges users may encounter and tips on how to overcome them:

Finding the Right Dosage

Determining the correct dosage can be tricky, especially since peptide dosages can vary depending on individual goals, body weight, and peptide type. Taking too much can lead to side effects, while taking too little might not yield the desired results.

Solution: Start with the lowest effective dose as recommended by your healthcare provider or peptide instructions in this book. Gradually increase the dosage if necessary while monitoring your body's response. Keep track of any side effects or improvements and consult with a healthcare professional if adjustments are needed.

Injection Fear or Discomfort

Many peptides are administered through subcutaneous injections, which can be intimidating or uncomfortable for those unfamiliar with needles.

Solution: Educate yourself on proper injection techniques or ask a healthcare professional to demonstrate. Use smaller, insulin-grade needles, and apply an ice pack to numb the area before injecting. Over time, the process becomes more routine and less intimidating.

Unregulated Peptide Market

Peptide quality can vary greatly depending on the supplier, especially in an unregulated market where some products might be counterfeit or contaminated.

Solution: Always purchase peptides from reputable sources that provide third-party testing or certificates of analysis (COAs). Stick with suppliers that have a good reputation in the peptide community and offer clear, transparent information about their products.

Slow or Inconsistent Results

Some users may become frustrated if they don't see immediate results. While peptides can offer significant benefits, the effects might take several weeks or even months to become noticeable.

Solution: Patience is key. Peptides work gradually, especially those targeting fat loss, muscle growth, or anti-aging effects. Stick with your regimen, track progress, and adjust as. If results seem stagnant, consult with a healthcare professional to discuss modifying your dosage or stack.

Cost of Peptide Therapy

Issue: Peptides can be expensive, especially when using multiple peptides in a stack or over long periods. For some users, the cost can be prohibitive.

Solution: Prioritize the peptides that align most closely with your goals. If cost is a concern, consider using fewer peptides but cycling them more strategically to still achieve results. Additionally, keep an eye out for reputable suppliers offering discounts for bulk purchases or loyalty programs.

Managing Side Effects

While peptides are generally well-tolerated, some individuals may experience mild side effects like headaches, nausea, or swelling at the injection site.

Solution: To minimize side effects, start with a low dose and increase gradually. Make sure you're following proper injection techniques and rotating injection sites to avoid irritation. If side effects persist, consult a healthcare provider to assess whether dosage adjustments or cycling off the peptide temporarily is necessary.

3.6 Common Mistakes to Avoid When Starting Peptides

Starting peptide therapy can be exciting, but there are a few common mistakes that beginners often make, which can impact the effectiveness of the treatment or lead to unnecessary side effects. Here are some pitfalls to avoid:

- **Incorrect Dosage:** One of the most frequent mistakes is taking too much or too little of the peptide. Always follow the dosage recommendations from your healthcare provider or the product's guidelines. Taking more than recommended won't necessarily speed up results and could lead to side effects such as headaches, fatigue, or nausea.
- **Poor Storage:** Peptides are sensitive to heat and light and must be stored correctly to maintain their potency. Always store peptides in a cool, dark place, and most should be refrigerated. If improperly stored, peptides can degrade, rendering them less effective or even useless.
- **Skipping Doses:** Consistency is key when using peptides. Skipping doses or not following the proper schedule can reduce the peptide's effectiveness. To get the best results, follow the recommended dosing schedule closely, and set reminders if needed.
- **Using Unreliable Sources:** Buying peptides from unverified or low-quality suppliers is a risky mistake. Always purchase peptides from reputable companies that provide third-party testing to ensure product purity and safety. Using low-quality or counterfeit peptides can lead to harmful side effects and wasted money.
- **Ignoring Cycling Guidelines:** Failing to cycle peptides properly can lead to reduced effectiveness and potential side effects over time. Always follow cycling recommendations and give your body time to reset between peptide cycles.

CHAPTER 4. SAFETY AND REGULATIONS

4.1 Peptide Safety: Understanding Side Effects and Risks

Peptide therapy is generally considered safe, especially when the peptides are sourced from reputable suppliers and administered correctly. However, like any treatment, peptides can have side effects, and it is important to understand the risks before starting therapy. Most people experience minimal or no side effects when using peptides, but individual reactions can vary based on factors such as dosage, method of administration, and the specific peptide being used.

Common Side Effects:

- **Injection Site Reactions:** The most common side effects are mild reactions at the injection site, such as redness, swelling, or irritation. These symptoms usually resolve quickly and are not cause for concern.

- **Headaches and Fatigue:** Some users report headaches or fatigue, especially when first starting peptide therapy or when taking higher doses. If this occurs, it's advisable to reduce the dosage and see if symptoms improve.

- **Nausea and Digestive Issues:** Certain peptides, particularly those affecting metabolism or appetite (like **Semaglutide**), can cause nausea or upset stomach. In most cases, these side effects diminish as your body adjusts to the peptide.

- **Hormonal Imbalances:** Peptides that influence hormone levels, such as growth hormone-releasing peptides, can cause temporary hormonal imbalances. This may result in symptoms such as water retention, joint pain, or increased hunger. If these symptoms are severe or persist, it's important to adjust the dosage or take a break from the peptide to allow the body to reset.

Less Common, But Serious Side Effects:

- **Hyperpigmentation:** Peptides like **Melanotan II**, which stimulate melanin production, may cause changes in skin pigmentation. While this effect is desired for tanning, in rare cases, it can lead to uneven skin tones or dark spots.

- **Excessive Growth Hormone Levels:** Overuse of growth hormone-releasing peptides can lead to excessive growth hormone levels, which may cause side effects such as increased blood sugar levels, carpal tunnel syndrome, or abnormal growth of tissues.

- **Allergic Reactions:** Although rare, some individuals may have an allergic reaction to peptides. Symptoms could include rashes, itching, or difficulty breathing. In such cases, discontinue use and seek medical attention immediately.

How to Minimize Risks:

- **Start with a Low Dose:** When beginning peptide therapy, always start with the lowest recommended dose and gradually increase as needed. This allows your body to adjust and reduces the risk of side effects.

- **Monitor Your Body:** Pay close attention to how your body reacts to the peptide. If you experience any side effects, consult with a healthcare provider, adjust your dosage, or consider cycling off the peptide temporarily.

- **Consult a Doctor or Healthcare Professional:** Before starting any peptide regimen, it's important to speak with a healthcare professional who can help guide you in choosing the right peptide, dosage, and method of administration.

4.2 Legal and Regulatory Considerations in Peptide Usage

The legal status of peptides varies depending on the country and the specific peptide in question. Some peptides are approved for medical use, while others are considered experimental or are not regulated, which creates a gray area when it comes to purchasing and using them.

Prescription vs. Over-the-Counter

In many countries, certain peptides, such as **insulin** or **growth hormone** (somatropin), are prescription-only medications. These peptides are regulated due to their powerful effects and potential for misuse. For example, growth hormone is a controlled substance in some countries because of its association with performance enhancement in sports. Other peptides, particularly newer or experimental ones, may not yet be approved for therapeutic use by regulatory bodies such as the **U.S. Food and Drug Administration (FDA)** or **European Medicines Agency (EMA)**.

Sports and Doping Regulations

Athletes need to be especially careful when using peptides, as many are banned by sports organizations such as the World Anti-Doping Agency (WADA). Peptides like IGF-1 LR3 or CJC-1295 are often prohibited in competitive sports because they can provide an unfair advantage by promoting muscle growth or enhancing recovery. If you are a competitive athlete, make sure to consult your sport's governing body or check WADA's list of banned substances to avoid penalties or disqualification.

Research Chemicals

Many peptides are sold online as research chemicals. This means they are legally available for purchase but are marketed for research purposes only, not for human use. This classification allows companies to sell peptides that have not been approved by regulatory authorities for medical or therapeutic use. While these peptides can still be effective and safe when used properly, purchasing them comes with the risk that the product may not meet stringent safety or purity standards.

4.3 Peptides and FDA: Current Status of Approval

The **U.S. Food and Drug Administration (FDA)** has approved a limited number of peptides for medical use, particularly for conditions such as diabetes, cancer, and hormonal deficiencies. However, many peptides available on the market today are not FDA-approved, meaning they have not undergone the rigorous clinical testing required to confirm their safety and efficacy for human use. Some of the peptides that have received FDA approval include Insulin, Liraglutide, Semaglutide and Bremelanotide (PT-141).

Experimental Peptides

Many peptides, including those used for anti-aging, muscle growth, and cognitive enhancement, remain unapproved by the FDA. This does not necessarily mean they are unsafe, but it does mean that they have not been evaluated in large-scale clinical trials to determine their long-term safety and effectiveness. Examples of unapproved peptides include BPC-157, TB-500, CJC-1295, and Ipamorelin.

CHAPTER 5. THERAPEUTIC PEPTIDES AND USES

5.1 Peptides for Fat Loss

Fat loss is one of the most sought-after benefits of peptide therapy, and there are several peptides specifically designed to help people lose fat while preserving lean muscle mass. Peptides used for fat loss typically work by increasing metabolism, reducing appetite, or improving the body's ability to break down and use stored fat.

Ipamorelin

Ipamorelin is a selective growth hormone-releasing peptide (GHRP) that has gained popularity for its ability to stimulate growth hormone (GH) production in the body. Ipamorelin helps promote lipolysis (the breakdown of fat) by increasing growth hormone secretion, which enhances metabolism and aids in reducing body fat. As a relatively mild peptide compared to other GHRPs, Ipamorelin offers a unique advantage: it triggers the release of growth hormone without significantly affecting other hormones like cortisol or prolactin. This makes it an excellent choice for individuals seeking muscle growth, fat loss, and recovery without the side effects of excessive hormone stimulation.

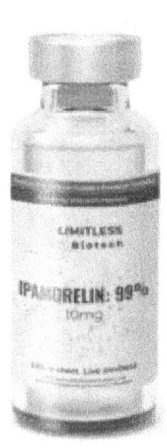

Benefits

Fat Loss: Ipamorelin helps increase lipolysis (fat breakdown) by promoting growth hormone release, making it easier for users to burn fat while preserving muscle.

Muscle Preservation: While promoting fat loss, Ipamorelin helps preserve lean muscle mass, which is often lost during dieting.

Improved Metabolism: Ipamorelin boosts metabolic rate, allowing the body to burn more calories even at rest, leading to sustained fat loss over time.

Method of Delivery

Ipamorelin is administered via subcutaneous injection, usually around the abdomen.

Recommended Dosage and Cycles

The standard dosage of Ipamorelin is between **200–300 mcg per injection**, administered 1–3 times daily. Most users begin with a lower dose and gradually increase based on their response to the peptide.

It is often used in cycles of **8–12 weeks**, followed by a break to avoid desensitization.

AOD-9604

AOD-9604 is a peptide that has shown significant potential in fat loss. It is a modified form of a specific region of the human growth hormone molecule responsible for fat metabolism. Unlike growth hormone,

AOD-9604 does not increase insulin resistance, making it a safer option for those with metabolic concerns. AOD-9604 works by mimicking the fat-burning effects of growth hormone without its adverse side effects, such as raising blood sugar levels. It has been used to help individuals lose weight, particularly in reducing body fat.

Benefits

- **Promotes Fat Breakdown**: AOD-9604 stimulates lipolysis, allowing the body to break down fat more effectively.

- **Does Not Affect Blood Sugar**: One of the key advantages of AOD-9604 is its ability to promote fat loss without affecting insulin or glucose metabolism, making it suitable for individuals with metabolic concerns like diabetes.

- **Improves Weight Loss**: Regular use of AOD-9604 can enhance overall weight loss, particularly in stubborn areas like the abdomen and thighs.

Method of Delivery and Dosage

AOD-9604 is administered via subcutaneous injection. The typical dosage for fat loss is **300 mcg per day,** and it can be used for 12–16 weeks in fat-loss cycles.

Semaglutide

Semaglutide, originally developed to treat type 2 diabetes, has gained attention for its powerful fat-loss effects. Semaglutide is a glucagon-like peptide-1 (GLP-1) receptor agonist that regulates insulin and glucose levels. However, one of its most significant benefits is appetite suppression. In clinical studies, Semaglutide has been shown to help individuals lose weight by reducing their appetite and improving their body's ability to process fats. This peptide has become a popular for weight loss, particularly for people struggling with obesity or those looking for a safe, non-invasive way to control their appetite and lose weight. Semaglutide works by slowing gastric emptying, making individuals feel fuller for longer, which leads to reduced calorie intake and weight loss.

Benefits

- **Appetite Suppression**: Semaglutide reduces hunger by slowing digestion, helping users naturally eat less without feeling deprived.

- **Improved Weight Loss**: Clinical trials have shown significant weight loss in individuals using Semaglutide, making it one of the most effective medications for weight reduction.

- **Blood Sugar Regulation**: In addition to promoting weight loss, Semaglutide helps regulate blood sugar levels, which can prevent spikes in glucose and insulin, making it particularly useful for individuals with insulin resistance.

Method of Delivery and Recommended Dosage

Semaglutide is administered via subcutaneous injection, typically once a week.

The starting dose is **0.25 mg per week**, gradually increasing to 1.0 mg per week as tolerated. For weight loss, treatment is usually continued for 16–24 weeks, or until the desired weight is achieved.

Tirzepatide

Tirzepatide, another GLP-1 receptor agonist, works similarly to Semaglutide but targets both GLP-1 and GIP (glucose-dependent insulinotropic polypeptide) receptors. This dual action makes Tirzepatide even more effective for fat loss. It improves the body's insulin sensitivity, helps regulate blood sugar levels, and significantly reduces appetite, leading to more profound fat loss than Semaglutide alone. Tirzepatide has become a highly sought-after peptide for people looking to lose significant amounts of weight while preserving muscle mass and improving overall metabolic health. It is one of the newer peptides used for obesity and metabolic health, offering superior appetite control and fat reduction.

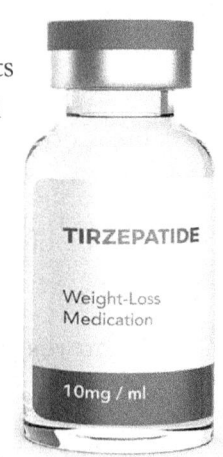

Benefits

Significant Fat Loss: Clinical studies have shown that Tirzepatide leads to greater fat loss compared to standard GLP-1 receptor agonists. It increases both fat oxidation and appetite suppression, promoting rapid and sustained weight reduction.

Improved Insulin Sensitivity: Tirzepatide improves insulin sensitivity, making it an ideal peptide for individuals with insulin resistance or type 2 diabetes.

Metabolic Health: Beyond weight loss, Tirzepatide supports overall metabolic health by lowering blood sugar levels, reducing cholesterol, and improving cardiovascular health.

Method of Delivery and Recommended Dosage

Tirzepatide is injected subcutaneously once per week, starting at **2.5 mg per week** and gradually increasing to **15 mg per week** based on tolerance and weight loss goals. It is typically used in cycles of **16–24 weeks** for significant fat loss.

Tesofensine

Tesofensine is a serotonin-noradrenaline-dopamine reuptake inhibitor (SNDRI) that was initially developed as a treatment for neurodegenerative diseases like Alzheimer's and Parkinson's. However, its potent appetite-suppressing properties led to its development as a weight loss agent. By increasing the levels of neurotransmitters like serotonin, dopamine, and norepinephrine, Tesofensine reduces appetite and increases metabolic rate, leading to weight loss.

Benefits

Appetite Suppression: Tesofensine's ability to increase serotonin and dopamine levels helps reduce hunger, making it easier to follow a calorie-restricted diet.

Fat Loss: By increasing metabolism and energy expenditure, Tesofensine helps the body burn more calories throughout the day, leading to fat loss.

Improved Mood and Motivation: The increase in dopamine levels can improve mood and motivation, which are often challenges during weight loss journeys.

Method of Delivery and Recommended Dosage

Tesofensine is taken orally, with the recommended dosage being **0.5 mg per day**. For weight loss, it is typically cycled for **12–16 weeks**, with users monitoring for any cardiovascular side effects, such as increased heart rate or blood pressure.

Tesamorelin

Tesamorelin is a growth hormone-releasing hormone (GHRH) analog that stimulates the pituitary gland to release more growth hormone. It has been used primarily to reduce visceral fat in individuals with HIV-associated lipodystrophy but has since gained popularity for its ability to reduce abdominal fat and preserve muscle mass in general populations.

Benefits

- **Reduction of Visceral Fat**: Tesamorelin specifically targets visceral fat, the fat stored around organs, which is particularly dangerous and difficult to lose. Studies show significant reductions in abdominal fat in individuals using Tesamorelin.
- **Muscle Preservation**: Tesamorelin helps preserve lean muscle mass during weight loss, which is often a concern for individuals trying to reduce fat without losing muscle.
- **Improved Metabolism**: By stimulating growth hormone release, Tesamorelin boosts metabolism, leading to fat loss while maintaining muscle mass.

Method of Delivery and Recommended Dosage

Tesamorelin is administered via subcutaneous injection, usually once daily. The typical dosage is **2 mg per day: 1 mg** at night, 90 minutes after last meal of the day and **1 mg** after waking up.

It is often cycled for **12–16 weeks**. Regular monitoring of blood sugar levels is recommended during use.

MOTS-C

MOTS-C is a mitochondrial-derived peptide that plays an important role in regulating metabolism and energy production. It improves the body's ability to burn fat by optimizing mitochondrial function, making it a powerful peptide for weight loss and improving metabolic health.

Benefits

Fat Oxidation: MOTS-C boosts mitochondrial function, which increases the body's ability to oxidize fat for energy. This leads to enhanced fat loss, particularly during exercise.

Improved Insulin Sensitivity: MOTS-C improves the body's response to insulin, making it easier to regulate blood sugar levels and reduce fat storage.

Increased Energy Levels: By improving mitochondrial efficiency, MOTS-C enhances overall energy levels, making it easier to stay active and maintain an exercise routine during weight loss.

Method of Delivery and Dosage

MOTS-C is administered via subcutaneous injection. The recommended dosage is **10 mg per week**, usually split into 2–3 injections. It is commonly used in fat loss cycles of **12–16 weeks** for best results.

5-Amino 1MQ

5-Amino 1MQ is a small molecule that inhibits the enzyme NNMT (nicotinamide N-methyltransferase), which plays a role in slowing metabolism. By inhibiting NNMT, 5-Amino 1MQ boosts cellular metabolism, leading to increased fat loss and enhanced energy levels.

Benefits

- **Fat Loss**: 5-Amino 1MQ helps increase metabolic rate by improving the body's ability to burn fat at the cellular level.

- **Improved Energy Levels**: Users often report increased energy and vitality due to the enhanced cellular function, making it easier to stay active during fat loss programs.

- **Preservation of Lean Mass**: While promoting fat loss, 5-Amino 1MQ helps preserve muscle mass, which is critical for maintaining a healthy body composition.

Side Effects

- Due to increased energy levels, some users may have difficulty sleeping if taken late in the day.

Method of Delivery and Dosage

5-Amino 1MQ is taken orally in capsule form, with the recommended dosage being **50–100 mg per day**, divided into two doses. It is typically cycled for **3–4 weeks** in fat loss programs, followed by a 1–2-week break.

5.2 Peptides for Muscle Growth and Performance

Peptides designed to improve muscle growth and performance are widely used by athletes, bodybuilders, and fitness enthusiasts. These peptides help increase muscle mass, speed up recovery, and improve overall athletic performance by stimulating growth hormone release, boosting protein synthesis, and reducing muscle breakdown.

Sermorelin

Sermorelin is a synthetic version of growth hormone-releasing hormone (GHRH), specifically designed to stimulate the pituitary gland to produce and release more growth hormone. Unlike synthetic human growth hormone (HGH), which introduces exogenous hormones into the body, Sermorelin encourages the body to increase its own production of growth hormone, leading to more natural and sustained effects.

Sermorelin is known for being a safer alternative to HGH therapy as it stimulates the body's natural hormonal pathways, reducing the risk of excessive growth hormone levels and associated side effects. The peptide is often used in anti-aging protocols as well as fitness and performance programs.

Benefits

Promotes Muscle Growth: By increasing growth hormone levels, Sermorelin improves muscle protein synthesis, allowing for faster muscle recovery and increased lean muscle mass.

Increased Recovery and Healing: Sermorelin can significantly speed up recovery times after intense workouts or injuries, allowing athletes to train more frequently without the risk of overtraining.

Fat Loss and Metabolism: Increased levels of growth hormone also promote lipolysis, the breakdown of fats. This makes Sermorelin a valuable tool for reducing body fat while maintaining or gaining lean muscle.

Improved Sleep and Recovery: Growth hormone peaks during deep sleep, and Sermorelin helps users achieve more restorative sleep, leading to better overall recovery and physical rejuvenation.

Method of Delivery and Recommended Dosage

Sermorelin is administered via subcutaneous injection, typically before bedtime to align with the body's natural growth hormone release cycles.

The typical dosage is **200–500 mcg per day,** depending on the user's goals and overall health. It is often cycled for **12–16 weeks**, followed by a break to prevent desensitization.

BPC-157

BPC-157 (Body Protection Compound 157) is a powerful peptide known for its ability to promote muscle and tissue repair. It is a peptide derived from a protein found in gastric juice.

While not directly linked to muscle growth, BPC-157 speeds up recovery from injuries and muscle damage, allowing athletes to return to their training more quickly. It works by promoting the healing of damaged tissues, improving blood flow to injured areas, and reducing inflammation. This makes BPC-157 particularly useful for anyone recovering from muscle tears, tendon injuries, or joint issues.

What makes BPC-157 particularly unique is its ability to increase blood flow to damaged areas, promote angiogenesis (the formation of new blood vessels), and accelerate the healing process in both acute and chronic injuries.

Benefits

Accelerated Muscle and Tissue Repair: BPC-157 stimulates the repair of damaged muscle fibers, tendons, and ligaments, significantly reducing recovery times for injuries.

Joint and Ligament Healing: In addition to muscle repair, BPC-157 promotes the healing of tendons and ligaments, which are notoriously slow to heal. This can help prevent chronic issues and improve joint mobility and flexibility.

Gut Health and Inflammation: BPC-157 was initially studied for its effects on gut health, particularly in healing ulcers and reducing inflammation in the digestive tract. Its anti-inflammatory properties extend to the entire body, making it useful in reducing chronic pain and inflammation in muscles and joints.

Improved Recovery from Workouts: By promoting faster tissue repair and reducing inflammation, BPC-157 allows users to recover more quickly from intense training sessions, enabling more frequent and productive workouts.

Method of Delivery and Recommended Dosage

BPC-157 is administered via subcutaneous injection, usually near the site of injury or discomfort. For systemic healing, injections can be made in the abdominal area. The typical dosage is **200–500 mcg per injection**, administered once or twice daily, depending on the severity of the injury and the user's goals.

Cycle Duration: BPC-157 can be used for periods of **4–12 weeks**, depending on the severity of the injury and the healing progress. Users should take a break after each cycle to avoid desensitization.

TB-500

TB-500 is a synthetic version of a naturally occurring peptide called Thymosin Beta-4, which is found in almost all human cells. Its primary function is to promote tissue repair and regeneration by increasing cell migration and differentiation. TB-500 is particularly well-known for its ability to heal injuries in muscles, tendons, ligaments, and even organs. It is commonly used in sports and fitness for its remarkable recovery-enhancing properties and its ability to reduce inflammation.

It plays an important role in angiogenesis (the formation of new blood vessels), wound healing, and reducing the buildup of scar tissue. This makes it especially valuable for athletes and individuals recovering from physical injuries, surgeries, or chronic inflammation. It also helps improve

flexibility and mobility by facilitating the healing of tendons and ligaments, which are slow to repair naturally.

Benefits

Accelerated Recovery from Injuries: TB-500 promotes faster healing by encouraging the migration of cells to the site of injury. It supports the repair of muscles, tendons, ligaments, and even the cardiovascular system. This helps in speeding up recovery times for both acute and chronic injuries.

Improved Flexibility and Mobility: TB-500 aids the healing of tendons and ligaments, which can lead to improved joint flexibility and range of motion.

Reduced Inflammation: TB-500 has potent anti-inflammatory properties that help reduce swelling, pain, and inflammation in both acute injuries and chronic conditions like arthritis. This allows users to heal more quickly and with less discomfort.

Cardiovascular Health: By promoting angiogenesis and tissue regeneration, TB-500 may also support cardiovascular health by improving blood flow and healing damaged blood vessels.

Method of Delivery

TB-500 is administered via subcutaneous injection, with users typically injecting the peptide near the site of injury for localized effects. For overall recovery, injections can be administered in the abdominal area.

Recommended Dosage and Cycles

The typical dosage of TB-500 ranges from **2–5 mg per week**, divided into **2–3 injections**. For users looking to accelerate recovery, the loading phase usually consists of **4–5 mg per week** for the first **4–6 weeks**.

- **Maintenance Phase**: After the initial loading phase, the dosage can be reduced to **2–3 mg per week** to maintain the peptide's effects and continue supporting recovery.

TB-500 cycles typically last between **4–8 weeks**, depending on the severity of the injury and the user's recovery needs.

IGF-1 LR3

IGF-1 LR3 (Insulin-like Growth Factor-1 Long R3) is a peptide that directly promotes muscle growth. IGF-1 is a hormone naturally produced by the liver in response to growth hormone stimulation. It is responsible for many of the growth hormone's anabolic effects, such as increasing protein synthesis and promoting muscle cell proliferation.

IGF-1 LR3 is a modified version of IGF-1 with a longer half-life, allowing it to remain active in the body for an extended period. This means users experience more sustained muscle growth and fat loss. Athletes and bodybuilders commonly use IGF-1 LR3 to build muscle mass, improve strength, and enhance overall physical performance. It also increases protein synthesis and promotes the uptake of amino acids into cells, further enhancing muscle growth and recovery.

Benefits

Muscle Growth and Hypertrophy: IGF-1 LR3 promotes significant muscle growth by increasing the size and number of muscle fibers. It activates satellite cells, which are essential for muscle repair and hypertrophy, making it a popular choice among bodybuilders and athletes looking to maximize muscle gains.

Enhanced Recovery: IGF-1 LR3 accelerates recovery by promoting protein synthesis and the repair of damaged tissues. This allows athletes to recover more quickly from intense training sessions, reducing downtime and the risk of injury.

Improved Strength and Performance: By increasing muscle mass and promoting recovery, IGF-1 LR3 enhances overall strength and athletic performance, making it ideal for strength training and competitive sports.

Fat Loss: IGF-1 LR3 also has fat-burning properties, as it increases metabolism and promotes the breakdown of fat stores for energy. This helps users achieve a leaner physique while building muscle.

Method of Delivery and Dosage

IGF-1 LR3 is typically administered via subcutaneous or intramuscular injection. Due to its longer half-life compared to regular IGF-1, fewer injections are required to maintain stable levels.

Dosage: The standard dosage ranges from **20–100 mcg per day**, with beginners starting at the lower end to assess tolerance. More experienced users may increase the dose as needed to promote greater muscle growth.

- **Cycle Duration**: IGF-1 LR3 is commonly cycled for **4–6 weeks**, followed by a break to avoid potential side effects and to allow the body's natural IGF-1 levels to return to normal.

DSIP

DSIP, or Delta Sleep-Inducing Peptide, is a neuropeptide known for its ability to promote restful sleep, particularly deep sleep, which is essential for recovery and tissue repair. Discovered in the 1970s, DSIP has gained attention for its potential in improving sleep quality, reducing stress, and supporting recovery in athletes and individuals with sleep disorders. Unlike traditional sleep aids, DSIP works by regulating the sleep-wake cycle and improving the body's natural sleep mechanisms, rather than sedating the user.

Sleep plays a vital role in recovery, especially for those engaged in intense physical training or recovering from injuries. DSIP's ability to promote deep, restful sleep makes it particularly valuable for athletes and individuals looking to optimize muscle recovery, growth, and overall health.

Benefits

Sleep Quality: DSIP promotes deeper, more restorative sleep by regulating the body's circadian rhythm and encouraging the onset of slow-wave sleep (deep sleep). This allows for recovery and reduces the risk of sleep disturbances.

Improved Recovery: Since the body releases the majority of growth hormone during deep sleep, DSIP indirectly enhances muscle recovery and growth by supporting better sleep cycles. This is especially beneficial for athletes who need optimal recovery after intense training.

Stress Reduction: DSIP has been shown to reduce stress and anxiety levels, which can interfere with sleep quality and recovery. By promoting relaxation, DSIP helps individuals fall asleep more easily and stay asleep longer.

Method of Delivery and Recommended Dosage

DSIP is administered via subcutaneous injection,

Dosage: The standard dosage of DSIP is **100–200 mcg per day**, administered approximately 30–60 minutes before bedtime.

- **Cycle Duration**: DSIP can be used on a continuous basis for several weeks or months, though it is often cycled for **4–6 weeks**.

GHRP-2

GHRP-2 (Growth Hormone Releasing Peptide-2) is a powerful growth hormone secretagogue that stimulates the pituitary gland to release more growth hormone (GH). It is one of the most potent GHRPs available and is widely used to promote muscle growth, fat loss, and recovery. GHRP-2 works by mimicking the effects of ghrelin, a hunger-stimulating hormone, and binding to specific receptors in the pituitary gland, leading to increased secretion of growth hormone.

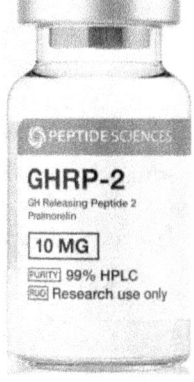

Benefits

Increased Growth Hormone Levels: GHRP-2 significantly boosts the release of growth hormone, which leads to muscle growth, improved recovery, and increased strength.

Muscle Growth and Recovery: Higher growth hormone levels promote muscle protein synthesis and tissue repair, allowing users to recover more quickly from intense training sessions and build lean muscle mass.

Fat Loss: GHRP-2 promotes fat breakdown by increasing the body's metabolic rate and encouraging the use of stored fat for energy. This makes it an effective peptide for improving body composition.

Improved Sleep: Like many growth hormone-releasing peptides, GHRP-2 enhances the quality of sleep, especially deep sleep, which is essential for muscle recovery and overall health.

Method of Delivery and Dosage

GHRP-2 is administered via subcutaneous injection, typically in the abdominal area. It can also be used in combination with other peptides like CJC-1295 to maximize growth hormone release.

- **Dosage**: The recommended dosage is **100–300 mcg per injection**, taken 1–3 times daily. For best results, injections should be taken on an empty stomach to avoid interference with insulin.
- **Cycle Duration**: GHRP-2 is commonly cycled for **8–12 weeks**, followed by a break.

GHRP-6

GHRP-6 (Growth Hormone Releasing Peptide-6) is another potent growth hormone secretagogue that stimulates the release of growth hormone from the pituitary gland. Like GHRP-2, GHRP-6 mimics the effects of ghrelin, a hormone that regulates hunger and stimulates GH release. However, GHRP-6 is often preferred for its stronger ability to increase appetite, making it a popular choice among individuals looking to gain muscle mass and improve recovery.

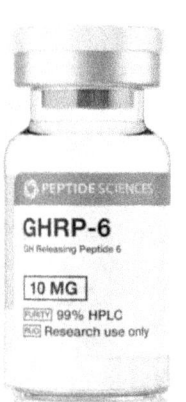

GHRP-6 is commonly used in muscle-building programs, as it promotes muscle growth, recovery, and supports fat loss.

Benefits

Increased Growth Hormone Release: GHRP-6 triggers a significant release of growth hormone, which promotes muscle growth, fat loss, and faster recovery.

Muscle Growth and Repair: By increasing growth hormone levels, GHRP-6 enhances muscle protein synthesis and tissue repair, allowing users to recover more quickly and build lean muscle mass.

Increased Appetite: One of the key benefits of GHRP-6 is its ability to stimulate appetite, making it ideal for individuals who struggle to consume enough calories for muscle growth.

Fat Loss: GHRP-6 promotes lipolysis (fat breakdown), making it a useful tool for improving body composition by reducing fat while increasing muscle mass.

Improved Sleep: GHRP-6 improves the quality of sleep, particularly deep sleep, which is essential for muscle recovery and overall health.

Method of Delivery and Dosage

GHRP-6 is administered via subcutaneous injection, usually 1–3 times per day. It is often used in combination with other peptides to enhance muscle growth and recovery.

Dosage: The typical dosage is **100–300 mcg per injection**, taken 1–3 times per day. For optimal results, GHRP-6 should be administered on an empty stomach, as food (especially carbohydrates and fats) can interfere with its effects.

Cycle Duration: GHRP-6 is typically cycled for **8–12 weeks**, followed by a break.

Hexarelin

Hexarelin is one of the most potent GHRPs available, known for its strong ability to stimulate growth hormone release. It is a synthetic peptide that mimics the effects of ghrelin and binds to specific receptors in the pituitary gland, causing a surge in growth hormone levels. Hexarelin is often used for muscle growth, recovery, and fat loss, but it also has unique benefits for cardiovascular health.

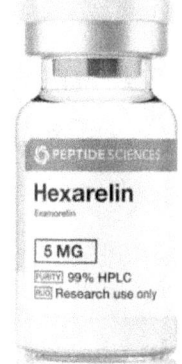

One of the distinguishing characteristics of Hexarelin is its potency as it can cause a more pronounced and sustained release of growth hormone compared to other GHRPs like GHRP-2 or GHRP-6. This makes it highly effective for individuals looking for rapid muscle growth and recovery, though it should be used with caution due to its potency.

Benefits

Significant Growth Hormone Release: Hexarelin is known for its powerful ability to stimulate growth hormone release, which leads to increased muscle growth, fat loss, and enhanced recovery.

Muscle Growth and Repair: By promoting higher levels of growth hormone, Hexarelin enhances muscle protein synthesis and accelerates tissue repair.

Fat Loss: Hexarelin promotes fat metabolism by increasing the body's metabolic rate and encouraging the breakdown of fat stores for energy. This helps users achieve a leaner physique while building muscle.

Improved Cardiovascular Health: Hexarelin has shown potential benefits for cardiovascular health by improving heart function and reducing the risk of heart-related issues. It promotes healing in heart tissue and may support recovery in individuals with cardiovascular conditions.

Method of Delivery and Dosage

Hexarelin is administered via subcutaneous injection. Due to its potency, lower doses are often recommended for those new to GHRPs.

Dosage: The recommended dosage is **100–200 mcg per injection**, taken 1–2 times per day. Due to its strong effects on growth hormone release, lower doses are often sufficient to achieve the desired results.

Cycle Duration: Hexarelin is typically cycled for **4–6 weeks** followed by a break.

PEG-MGF

PEG-MGF (Pegylated Mechano Growth Factor) is a modified version of IGF-1 that is primarily responsible for repairing and regenerating muscle tissues after intense exercise. Mechano Growth Factor (MGF) is naturally produced in the body in response to muscle damage or mechanical overload (such as weightlifting). PEG-MGF is a pegylated form of MGF, meaning it has been modified to have a longer half-life, allowing it to stay active in the bloodstream for longer and promote more sustained muscle growth and repair.

Benefits

Muscle Repair: PEG-MGF promotes the repair and regeneration of muscle tissues after exercise-induced damage, allowing for faster recovery and more significant muscle growth.

Muscle Growth and Hypertrophy: By activating satellite cells (the precursors to muscle cells), PEG-MGF encourages the growth of new muscle fibers, leading to increased muscle size and strength.

Improved Recovery: PEG-MGF reduces recovery times after intense training sessions, allowing athletes to train more frequently without overtraining or risking injury.

Longer Half-Life: The longer the half-life of MGF allows for more sustained effects on muscle growth and recovery.

Method of Delivery and Dosage

PEG-MGF is typically administered via subcutaneous or intramuscular injection, depending on the user's preference and goals.

Dosage: The recommended dosage for PEG-MGF is **200–400 mcg per injection**, taken **2–3 times per week**. It is often injected post-workout to maximize its effects on muscle repair and recovery.

Cycle Duration: PEG-MGF is commonly cycled for **4–6 weeks**, with a break.

MK-677

MK-677, also known as Ibutamoren, is an orally active growth hormone secretagogue that mimics the action of ghrelin, a hunger hormone, and stimulates the release of growth hormone (GH) and insulin-like growth factor 1 (IGF-1). Unlike many other peptides that require injections, MK-677 offers the convenience of oral administration.

MK-677 is unique in that it stimulates growth hormone release without significantly affecting cortisol or other stress hormones, making it a safer and more balanced option for long-term use. Its ability to maintain consistent growth hormone levels for 24 hours after a single dose makes it highly effective for muscle building and at reduction.

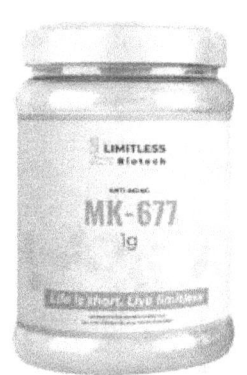

Benefits

Muscle Growth and Hypertrophy: MK-677 increases the release of growth hormone and IGF-1, both of which are needed for muscle protein synthesis and muscle hypertrophy.

Fat Loss: By promoting the breakdown of stored fat for energy (lipolysis) and increasing metabolic rate, MK-677 helps reduce body fat while preserving muscle mass. Its ability to improve body composition makes it popular for both bulking and cutting phases.

Recovery: Growth hormone plays a key role in tissue repair and recovery. MK-677 aids recovery from intense workouts by speeding up muscle repair, reducing muscle soreness, and improving overall recovery time.

Improved Bone Density: MK-677 has been shown to increase bone density, which is important for athletes and aging individuals looking to maintain strong and healthy bones.

Increased Appetite: Due to its ghrelin-mimicking effects, MK-677 increases appetite, which can be beneficial for individuals trying to consume more calories for muscle growth.

Method of Delivery and Dosage

MK-677 is taken orally, typically in the form of capsules or tablets. This makes it one of the most convenient peptides for users who prefer to avoid injections.

Dosage: The recommended dosage of MK-677 is **10–25 mg per day**. Beginners typically start with a lower dose (10 mg) and gradually increase based on their tolerance and desired effects.

Cycle Duration: MK-677 is often used for **8–12 weeks**, though some users extend their cycles to **16 weeks** for more significant muscle growth and fat loss.

Ipamorelin

Ipamorelin is a selective growth hormone-releasing peptide (GHRP) that stimulates the release of growth hormone from the pituitary gland without significantly affecting other hormones like cortisol or prolactin. It is one of the mildest and most well-tolerated GHRPs, making it a popular choice for individuals seeking to increase growth hormone levels for muscle growth, fat loss, and enhanced recovery with minimal side effects.

Unlike some other GHRPs that can lead to spikes in stress hormones or hunger, Ipamorelin provides a more targeted and controlled release of growth hormone. This makes it especially valuable for athletes and individuals looking for gradual, sustained improvements in muscle growth and recovery without the risk of hormonal imbalances.

Benefits

Muscle Growth and Recovery: Ipamorelin promotes muscle protein synthesis and aids tissue repair by increasing growth hormone levels.

Fat Loss: Growth hormone plays a key role in fat metabolism, and Ipamorelin enhances lipolysis (fat breakdown) by stimulating the release of growth hormone. This leads to improved body composition, with reductions in body fat and preservation of lean muscle mass.

No Impact on Cortisol or Prolactin: One of the main advantages of Ipamorelin over other GHRPs is its lack of significant effect on cortisol and prolactin levels, which means fewer side effects like increased stress or unwanted hormonal fluctuations.

Method of Delivery and Recommended Dosage

Ipamorelin is administered via subcutaneous injection, typically in the abdominal area.

Dosage: The standard dosage of Ipamorelin is **200–300 mcg per injection**, taken **1–3 times per day**. For most users, starting with one daily injection is sufficient, with higher doses reserved for individuals seeking more pronounced growth hormone release.

Cycle Duration: Ipamorelin is commonly used in cycles of **8–12 weeks**, followed by a break.

CJC-1295

CJC-1295 is a long-acting growth hormone-releasing hormone (GHRH) analog that stimulates the release of growth hormone from the pituitary gland. It is known for its ability to provide sustained growth hormone release over time, making it a powerful peptide for muscle growth, fat loss, and anti-aging benefits. The peptide's long half-life means that users can experience continuous growth hormone release without frequent injections, making it a convenient option for long-term use.

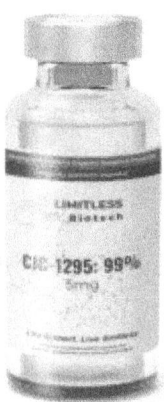

Benefits

Growth Hormone Release: CJC-1295 provides a prolonged release of growth hormone over several days, reducing the need for frequent injections. This sustained release promotes muscle growth, fat metabolism, and overall physical recovery.

Muscle Growth: By increasing growth hormone levels, CJC-1295 helps stimulate muscle protein synthesis and tissue repair, making it a popular choice for bodybuilders and athletes seeking to enhance muscle mass and recovery.

Fat Loss: Growth hormone plays a key role in fat metabolism, and CJC-1295 supports fat loss by promoting lipolysis. Users often report a decrease in body fat, particularly in stubborn areas like the abdomen and thighs.

Anti-Aging: CJC-1295's ability to increase growth hormone levels helps reduce the visible signs of aging, such as wrinkles and sagging skin. It also supports collagen production, which improves skin elasticity and overall skin health.

Recommended Dosage

CJC-1295 is administered via **subcutaneous** injection.

The standard dosage for CJC-1295 is **100-200 mcg (1 mg)** per injection, administered **1–2 times per week**. The peptide is often used in cycles of 8–12 weeks, followed by a break.

5.3 Peptides for Brain Health and Cognitive Performance

Brain health and cognitive performance has become an increasingly popular area of research in peptide therapy, as many people look for ways to boost memory, focus, and overall brain function. Peptides in this category are designed to improve mental clarity, support neuron health, and improve cognitive performance, making them useful for everyone from students and professionals to older adults concerned about cognitive decline.

Semax

Semax is a synthetic peptide derived from adrenocorticotropic hormone (ACTH) but without any hormonal activity. Developed in Russia in the 1980s for its neuroprotective and cognitive-enhancing properties, Semax has gained popularity for its ability to improve brain function, enhance memory, and promote neuroplasticity.

It is widely used for cognitive enhancement, mood regulation, and in the treatment of various neurological conditions. It has also been used to treat conditions like ADHD and depression, thanks to its neuroprotective properties and ability to regulate dopamine levels.

Semax is considered a nootropic, meaning it improves cognitive function, particularly in areas such as memory, learning, and mental clarity. It is also noted for its ability to increase the production of brain-derived neurotrophic factor (BDNF), a protein which supports the growth, development, and maintenance of neurons.

Benefits

Cognitive Performance: Semax is known for improving memory retention, learning abilities, and overall mental clarity. It improves cognitive performance in both healthy individuals and those suffering from cognitive decline.

Neuroprotection: By increasing BDNF levels, Semax supports the health and growth of neurons, protecting the brain from damage caused by stress, toxins, or neurological conditions.

Mood Regulation: Semax has been shown to regulate mood and reduce symptoms of anxiety and depression. It promotes a sense of well-being and emotional stability by modulating the levels of dopamine and serotonin in the brain.

Focus and Alertness: Users often report improved focus, attention, and mental energy when using Semax, making it an ideal peptide for individuals needing to stay alert and sharp for extended periods.

Method of Delivery and Recommended Dosage

Semax is most commonly administered intranasally, which allows for rapid absorption into the brain. It can also be injected subcutaneously, though nasal delivery is preferred for cognitive benefits.

Dosage: The typical nasal dosage of Semax is **100–300 mcg per spray**, used **1–2 times daily**. One spray in each nostril once or twice per day is often sufficient.

You will need 300 mcg of Semax per spray if your bottle has 30 mg of Semax in a 10 mL solution.

Dosage of **100-300 mcg** once a day if injected **subcutaneously**.

Cycle Duration: Semax can be used continuously for **2–4 weeks**, followed by a break. It may also be used intermittently, depending on the user's cognitive or mood needs.

Selank

Selank is a synthetic peptide derived from the naturally occurring peptide tuftsin, which plays a role in immune function. Developed in Russia, Selank is primarily used for its anxiolytic (anti-anxiety) and cognitive-enhancing properties. It has been shown to reduce anxiety, improve mood, and improve cognitive performance without causing the sedation or dependency associated with traditional anti-anxiety medications.

Selank modulates the levels of neurotransmitters in the brain, particularly serotonin, dopamine, and norepinephrine, all of which are involved in regulating mood, stress, and cognitive function. This makes it a valuable peptide for individuals seeking to improve mental clarity, reduce anxiety, and improve their overall sense of well-being.

Benefits

Reduces Anxiety: Selank is highly effective at reducing symptoms of anxiety and promoting emotional stability without the sedative effects of traditional anti-anxiety medications. It calms the mind while allowing users to remain alert and focused.

Cognitive Function: In addition to its anxiolytic properties, Selank enhances cognitive performance, particularly in the areas of memory, learning, and focus. It is often used by individuals seeking to improve mental clarity and cognitive stamina.

Mood Stabilization: Selank has been shown to stabilize mood and reduce symptoms of depression. By regulating serotonin and dopamine levels, it promotes a sense of calm and emotional balance.

Immune System Support: Interestingly, Selank also has immunomodulatory effects, supporting the immune system and helping the body respond more effectively to stress.

Risks and Side Effects

Selank is well-tolerated and has a low risk of side effects, making it an attractive option for individuals seeking natural anxiolytic and cognitive-enhancing solutions. However, some users may experience:

- **Nasal Irritation**: When used intranasally, mild irritation or discomfort in the nasal passages may occur.
- **Drowsiness**: In rare cases, some users may feel slightly drowsy, particularly when using higher doses of Selank.

Method of Delivery and Dosage

Selank is typically administered **intranasally**, allowing for fast absorption into the bloodstream and brain. It can also be administered via subcutaneous injection, though the nasal spray is the preferred method.

Dosage: The typical nasal dosage of Semax is **250–500 mcg per spray**, used **1–3 times daily**. One spray in each nostril, once or twice per day is often sufficient.

Dosage of **100-300 mcg** once a day if injected **subcutaneously**.

Cycle Duration: Selank can be used continuously for **4–6 weeks**, though many users prefer to use it on an as-needed basis for anxiety relief or cognitive support.

Dihexa

Dihexa is another peptide gaining attention for its potential in promoting brain health. Dihexa is a neuropeptide that can cross the blood-brain barrier, allowing it to directly influence brain function. It is known to promote the growth of new synapses, the connections between neurons, which are critical for learning and memory. Dihexa's ability to aid synaptic formation makes it particularly useful for individuals looking to improve cognitive performance or prevent cognitive decline associated with aging or neurodegenerative diseases.

Method of Delivery and Dosage

Dehexa is commonly administered via transdermal application.

Dosage: The typical dosage of Dihexa is **8–40 mg** used **once daily**.

Cerebrolysin

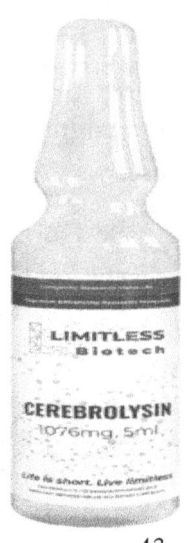

Cerebrolysin is a peptide mixture that contains neurotrophic factors known to stimulate neuron growth and protect against brain cell damage. It has been used in Europe to treat Alzheimer's disease, stroke, traumatic brain injuries, and cognitive decline. Cerebrolysin works by promoting the repair and regeneration of brain cells, improving cognitive function, and slowing down the progression of neurodegenerative diseases. It is particularly useful for older adults looking to preserve their cognitive abilities and maintain mental sharpness as they age.

Cerebrolysin's ability to cross the blood-brain barrier makes it particularly effective for enhancing brain function and promoting recovery from brain injuries or neurodegenerative conditions. It is used extensively in Europe and Asia, particularly in clinical settings for its potent cognitive and neurological benefits.

Benefits

Cognitive Perfomance: Cerebrolysin improves cognitive function, particularly in areas such as memory, learning, and mental clarity. It is commonly used to improve cognitive performance in both healthy individuals and those with cognitive impairments.

Neuroprotection: One of Cerebrolysin's key benefits is its ability to protect neurons from damage caused by oxidative stress, inflammation, and neurotoxins. This makes it highly effective in treating neurodegenerative conditions like Alzheimer's and Parkinson's disease.

Neuroplasticity and Recovery: Cerebrolysin promotes neuroplasticity, the brain's ability to form new neural connections. This is particularly beneficial for individuals recovering from stroke, traumatic brain injuries, or other neurological conditions.

Mood Stabilization and Cognitive Clarity: Some users report improvements in mood and emotional stability when using Cerebrolysin, likely due to its positive effects on brain function and neurochemical balance.

Method of Delivery and Dosage

Cerebrolysin is typically administered via intramuscular or intravenous injection. Its administration method and dosage depend on the severity of the condition being treated, as well as the user's cognitive goals.

Dosage: The standard dosage of Cerebrolysin ranges from **5–10 ml per day**.

For **Traumatic Brain Injury, 20–40 ml per day** is often used.

For **Alzheimer's Disease, 20–40 ml per day** is often used

For **Vascular Dementia, 20–40 ml per day** is often used.

For **Stroke, 20–40 ml per day** is often used.

For **cognitive enhancement** or neuroprotection, smaller doses of **5 ml** daily or every other day are often used.

Cycle Duration: Cerebrolysin is typically used in cycles of **10–20 days**, followed by a break. For more severe conditions, longer treatment cycles may be recommended under medical supervision.

Orexin A

Orexin A, also known as hypocretin-1, is a neuropeptide that helps in regulating wakefulness, arousal, and energy expenditure. It is produced in the hypothalamus and is responsible for maintaining wakefulness and preventing sleep. Orexin A has been studied for its potential to treat conditions such as narcolepsy and excessive daytime sleepiness, and it is also of interest for its potential to enhance cognitive function, improve focus, and increase alertness.

Orexin A is gaining attention as a potential cognitive enhancer due to its ability to improve mental alertness and energy levels without the jitteriness or dependency associated with traditional stimulants like caffeine or amphetamines.

Benefits

Wakefulness: Orexin A promotes wakefulness and reduces feelings of fatigue, making it ideal for individuals dealing with excessive daytime sleepiness or conditions like narcolepsy.

Cognitive Performance: By improving alertness and focus, Orexin A enhances cognitive function, particularly in tasks requiring sustained attention and mental clarity.

Energy and Mood: Orexin A is involved in regulating energy expenditure and mood, making it beneficial for individuals seeking to improve both physical and mental energy levels.

Appetite Regulation: Orexin A also plays a role in appetite regulation, helping to balance hunger and energy expenditure.

Method of Delivery and Dosage

Orexin A is typically administered **intranasally,** allowing for rapid absorption and immediate effects on wakefulness and alertness.

Dosage: The typical dosage of Orexin A is **100-150 mg per dose**, used **once per day,** usually in the morning.

PE-22-28

PE-22-28 is a synthetic peptide derived from the naturally occurring peptide Spadin, which is known to modulate the TREK-1 potassium channel in the brain. By blocking this channel, **PE-22-28** promotes neuroprotection and stress resilience, making it a valuable tool for individuals dealing with stress, anxiety, or cognitive decline. It has also been studied for its antidepressant effects and its potential to improve mood and cognitive function.

PE-22-28 works by promoting neurogenesis (the formation of new neurons) and protecting the brain from the damaging effects of chronic stress and neuroinflammation. This makes it particularly useful for individuals looking to improve their mental health, cognitive performance, and overall brain function.

Benefits

Neuroprotection: PE-22-28 promotes the health and survival of neurons, protecting the brain from damage caused by stress, inflammation, or neurotoxins.

Stress Resilience: By modulating the TREK-1 potassium channel, PE-22-28 improves the brain's ability to cope with stress, reducing symptoms of anxiety and promoting emotional resilience.

Cognitive Performance: PE-22-28 has been shown to improve cognitive function, particularly in areas such as memory, learning, and mental clarity.

Antidepressant Effects: Some studies suggest that PE-22-28 has antidepressant properties, making it a potential treatment for mood disorders and depression.

Method of Delivery and Dosage

PE-22-28 is typically administered via intranasally.

Dosage: The standard dosage is **400 mcg**, administered via nasal spray **once daily**, preferably in the morning.

Cycle Duration: PE-22-28 is typically used in cycles of **4–6 weeks**, followed by a break.

FGL

FGL (Fibroblast Growth Factor-Like Peptide) is a synthetic peptide designed to mimic the effects of the natural fibroblast growth factor (FGF) involved in promoting neuroplasticity and cognitive function. FGL has been studied for its ability to enhance memory, learning, and overall brain function by promoting the formation of new synaptic connections between neurons. It is of particular interest for its potential to treat neurodegenerative conditions, such as Alzheimer's disease, and to improve cognitive performance in healthy individuals.

By improving neuroplasticity, FGL supports the brain's ability to adapt, learn, and recover from injuries or cognitive decline.

Benefits

Memory and Learning: FGL promotes the formation of new neural connections, improving memory retention and learning capabilities. It is particularly effective for individuals looking to improve cognitive performance or recover from brain injuries.

Neuroplasticity: FGL supports the brain's natural ability to form new synapses, which is critical for learning, memory, and recovery from neurodegenerative conditions.

Neuroprotection: FGL protects neurons from damage caused by inflammation, oxidative stress, or neurotoxins, making it valuable for preventing cognitive decline or neurodegenerative diseases.

Cognitive Performance: By improving brain function, FGL improves mental clarity, focus, and overall cognitive performance.

Method of Delivery and Dosage

FGL is typically administered via subcutaneous injection.

Dosage: The standard dosage of FGL is **100–500 mcg per injection**, taken **1–2 times per day**.

Cycle Duration: FGL is typically used in cycles of **4–8 weeks**, depending on the user's cognitive goals and response to the peptide.

5.4 Peptides for Longevity and Anti-Aging

As people seek ways to live healthier, longer lives, peptides have emerged as a promising tool for slowing down the effects of aging and even reversing some of the damage that comes with it. As we age, the body's production of key peptides declines, leading to slower healing, decreased energy, and the breakdown of tissues. Peptides used in anti-aging therapies help address these issues by replenishing the body's natural supply and improving cellular and immune function.

Epitalon

One of the most promising peptides with anti-aging properties is **Epitalon,** also known as **Epithalon.** It works by stimulating the production of an enzyme called telomerase, which helps maintain the length of telomeres. Telomeres are protective caps at the ends of chromosomes that shorten as we age. Shortened telomeres are linked to aging and age-related diseases. By promoting telomere lengthening, **Epitalon** has the potential to slow down aging at the cellular level, which can lead to improvements in overall vitality, skin health, and even longevity. It also regulates the sleep-wake cycle by improving melatonin production, which becomes impaired with age.

Epitalon has gained popularity for its ability to promote cellular repair, boost immune function, regulate circadian rhythms, and slow the aging process at the cellular level.

Benefits

Telomere Extension: The most significant benefit of Epitalon is its ability to activate telomerase, which lengthens telomeres and protects cells from aging. Longer telomeres are associated with increased cellular lifespan and overall longevity.

Anti-Aging: Epitalon helps delay the aging process by promoting cellular repair and regeneration. It improves the function of key organs, boosts immune health, and enhances the body's ability to maintain homeostasis as it ages.

Improved Sleep and Circadian Rhythm: Epitalon has been shown to regulate melatonin production, helping to normalize sleep cycles and improve sleep quality, particularly in older adults.

Immune Function: Epitalon boosts immune system function by stimulating the activity of the pineal gland, which helps regulate the body's defense mechanisms. This improved immune function can help protect against age-related diseases and infections.

Potential Cancer Protection: Some research suggests that Epitalon may reduce the risk of cancer by protecting DNA from damage and supporting the body's natural tumor suppression mechanisms.

Method of Delivery and Dosage

Epitalon is most commonly administered via subcutaneous injection, though it can also be taken orally. However, **injectable forms** are generally considered more effective as oral formulations would break the peptide down.

Dosage: The typical dosage of Epitalon for anti-aging is **1–3 mg per day**, administered for **10–20 days**. This cycle can be repeated every **6–12 months**, depending on the user's goals and health status.

Ben Greenfield recommends **10 mg** of **Epitalon** injected subcutaneously three times a week for three weeks straight and **once a year**.

Cycle Duration: Epitalon is usually taken in short cycles, typically once or twice per year. Each cycle lasts **10–20 days**, with a break in between to prevent desensitization and maintain long-term effectiveness.

Thymalin

Thymalin is another peptide used to promote longevity by boosting immune system function. Thymalin is a thymic peptide derived from the thymus gland, an organ that plays a key role in regulating the immune system. As we age, the thymus gland shrinks, leading to a decline in immune function. Thymalin works by stimulating the production and activity of T-cells, which are essential for a healthy immune response and fighting off infections. This makes it an important peptide for both immune support and anti-aging purposes.

In addition to boosting the immune system, Thymalin has been shown to promote tissue repair, reduce inflammation, and support overall longevity.

Benefits

Boosted Immune Function: Thymalin improves the production and activity of T-cells, which are essential for fighting infections, viruses, and age-related immune decline. This immune support helps protect against age-related diseases and improves the body's ability to heal and repair itself.

Anti-Aging Effects: By promoting immune health and reducing inflammation, Thymalin helps delay the aging process at the cellular level. It improves the body's resilience to stress, supports tissue regeneration, and helps maintain youthful vitality.

Reduced Inflammation: Thymalin has potent anti-inflammatory properties, helping to reduce chronic inflammation that can accelerate the aging process and contribute to age-related diseases such as arthritis, cardiovascular disease, and neurodegenerative disorders.

Tissue Repair and Regeneration: Thymalin promotes the repair of damaged tissues and accelerates wound healing, making it valuable for individuals recovering from injuries or surgeries, particularly in older adults.

Method of Delivery and Dosage

Thymalin is typically administered via subcutaneous injection, often in combination with other peptides such as Epitalon for enhanced anti-aging benefits.

Dosage: The standard dosage of Thymalin for immune support and anti-aging is **10–20 mg per day**, taken for **5–10 days**. This cycle can be repeated every **4–6 months**, depending on the user's health status and goals.

Cycle Duration: Thymalin is commonly used in short cycles of **5–10 days**, repeated every few months to maintain immune health and anti-aging benefits.

Recommended doasage for combination of Epitalon and Thymalin: 5 mg of Thymalin and Epitalon respectively, once per day for 20 days straight, repeated every 6 months.

GHK-Cu

GHK-Cu (Copper Peptide) is a naturally occurring peptide that plays a vital role in wound healing, tissue repair, and skin health. It was first discovered in the 1970s and has since become widely known for its regenerative properties, particularly in the areas of skin rejuvenation, anti-aging, and cellular repair. GHK-Cu promotes collagen production, reduces inflammation, and enhances cellular communication, making it a key peptide for improving skin elasticity, reducing wrinkles, and supporting overall cellular health.

GHK-Cu is often used in cosmetic products for its skin-rejuvenating effects, but its benefits go far beyond skincare. It has been shown to promote tissue regeneration, improve immune function, and even protect DNA from damage, making it a powerful tool for longevity and anti-aging.

Benefits

Skin Rejuvenation: GHK-Cu is known for its ability to promote collagen production, improve skin elasticity, and reduce the appearance of fine lines and wrinkles. It also helps fade scars and hyperpigmentation, making it popular in anti-aging skincare routines.

Wound Healing and Tissue Repair: GHK-Cu accelerates wound healing by promoting tissue regeneration and reducing inflammation. It aids the body's ability to repair damaged tissues, making it valuable for individuals recovering from injuries or surgeries.

Anti-Inflammatory Effects: GHK-Cu has potent anti-inflammatory properties, helping to reduce chronic inflammation that contributes to aging and age-related diseases.

Cellular Repair and DNA Protection: GHK-Cu protects cells from oxidative damage and promotes the repair of damaged DNA, which helps delay the aging process at the cellular level. This makes it a key player in longevity and anti-aging protocols.

Hair Growth: GHK-Cu has also been shown to promote hair growth by stimulating the hair follicles and improving scalp health, making it valuable for individuals dealing with hair loss or thinning hair.

Method of Delivery and Dosage

GHK-Cu can be administered in several forms, including topical creams, serums, and subcutaneous injections. Topical forms are typically used for skin rejuvenation, while injectable forms are used for systemic benefits such as tissue repair and cellular regeneration.

Topical Dosage: When used topically, GHK-Cu is typically applied at concentrations of **0.5%–1%** in serums or creams, applied once or twice daily for skin rejuvenation.

Injectable Dosage: When administered via injection, the standard dosage of GHK-Cu is **2–5 mg per injection**, taken once daily for **4–6 weeks**, depending on the desired effects.

Cycle Duration: For anti-aging and skin rejuvenation purposes, GHK-Cu can be used continuously in topical forms, while injectable forms are typically cycled for **4–6 weeks**, followed by a break.

Humanin

Humanin is a small mitochondrial-derived peptide that was first discovered in human brain cells. It has garnered attention for its ability to protect cells from oxidative stress, inflammation, and apoptosis (cell death), all of which are major contributors to the aging process. Humanin plays an important role in mitochondrial health, which is essential for energy production, cellular repair, and overall longevity.

Mitochondria, often referred to as the "powerhouses" of the cell, play a crucial role in aging. As we age, mitochondrial function declines, leading to reduced energy levels, increased cellular damage, and the development of age-related diseases. Humanin helps combat these effects by improving mitochondrial function, protecting cells from damage, and promoting overall longevity.

Benefits

Mitochondrial Protection: Humanin helps protect mitochondria from oxidative stress, reducing cellular damage and promoting energy production. This helps in improving cellular health and delaying the aging process.

Neuroprotection: Humanin has been shown to protect neurons from damage caused by oxidative stress, inflammation, and neurotoxins. This makes it particularly valuable for individuals seeking to prevent or slow down neurodegenerative conditions like Alzheimer's and Parkinson's disease.

Improved Longevity: By promoting mitochondrial health and protecting cells from damage, Humanin has the potential to increase lifespan and healthspan, allowing individuals to live longer, healthier lives.

Reduced Inflammation: Humanin has anti-inflammatory properties that help reduce chronic inflammation, a key driver of aging and age-related diseases.

Recommended Dosage

Humanin is commonly administered via **subcutaneous injection.**

Dosage: The standard dosage of Humanin is **1–5 mg per injection**, administered **once daily or every other day**. For neuroprotection and mitochondrial health, lower doses are often used for longer periods.

Cycle Duration: Humanin can be used in cycles of **4–6 weeks**, followed by a break to assess the user's response and adjust dosage as needed.

TB-4/TB-500

Thymosin Beta-4 is a synthetic version of a naturally occurring peptide found in nearly all human cells. It is known for its powerful ability to promote cell proliferation and migration, tissue repair, reduce inflammation, and enhance cellular regeneration. While it is widely used for its healing properties in muscle, tendon, and ligament injuries, it also plays a key role in supporting longevity by promoting overall tissue health and reducing age-related inflammation.

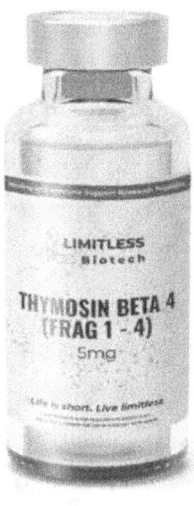

It is particularly valuable for older individuals or athletes recovering from injuries, as it accelerates the healing process, improves joint mobility, and supports long-term tissue health. Its systemic effects on tissue repair and regeneration make it a cornerstone peptide for those seeking to improve both performance and longevity.

Benefits

Improved Flexibility and Mobility: By promoting tissue repair and reducing inflammation, It improves joint flexibility and mobility, which is particularly beneficial for individuals dealing with age-related stiffness or joint pain.

Tissue Repair: It promotes the migration of cells to the site of injury, accelerating the healing of muscles, tendons, ligaments, and even organs. This makes it invaluable for individuals recovering from injuries or surgeries, particularly older adults.

Reduced Inflammation: It has strong anti-inflammatory properties that help reduce chronic inflammation, which can contribute to the aging process and age-related diseases such as arthritis and cardiovascular disease.

Longevity Support: It's ability to promote tissue repair and reduce systemic inflammation helps support long-term health and vitality, making it an important peptide in anti-aging and longevity protocols.

Method of Delivery and Dosage

It is typically administered via **subcutaneous injection.**

Dosage: The standard dosage ranges from **2–5 mg per week**, divided into 2–3 injections. For individuals recovering from injuries or seeking anti-aging benefits, a lower maintenance dose is often used after the initial healing phase.

Cycle Duration: It is commonly used in cycles of **4–8 weeks** for tissue repair, with a maintenance phase for ongoing support of joint health and longevity.

5.5 Peptides for Sexual Health

Peptides have shown significant potential in enhancing sexual health for both men and women. By addressing issues such as low libido, erectile dysfunction, and overall sexual performance, these peptides provide a natural and targeted approach to improving sexual well-being without the side effects associated with some traditional treatments.

PT-141

PT-141, also known as **Bremelanotide**, is a peptide derived from the melanocortin hormone. It was originally developed for its tanning properties but was soon found to have a potent effect on sexual arousal and desire. **PT-141** works by stimulating melanocortin receptors in the brain, which are involved in sexual arousal and desire.

Unlike drugs like Viagra, which target blood flow, PT-141 directly influences sexual desire, making it effective for both men and women. In men, it helps treat erectile dysfunction, while in women, it has been shown to increase sexual desire and arousal. PT-141 is particularly useful for people who have not responded well to other treatments or who experience low libido due to hormonal imbalances, stress, or age.

PT-141's approach to improving sexual health unique, as it not only helps with physical performance (such as erectile function in men) but also increases libido and sexual desire. It is effective for both men and women, making it a versatile option for addressing sexual dysfunction.

Benefits

Increased Sexual Desire: PT-141 stimulates sexual arousal and desire in both men and women by acting on melanocortin receptors in the brain. Users often report heightened libido and a stronger sexual response after taking PT-141.

Improved Erectile Function: For men, PT-141 has been shown to improve erectile function, particularly in cases where traditional ED medications have not been effective. By improving sexual arousal, PT-141 helps men achieve and maintain erections.

Enhanced Sexual Satisfaction for Women: PT-141 is one of the few peptides that have been studied specifically for its effects on female sexual health. It can improve sexual satisfaction, arousal, and orgasmic function in women, making it a valuable option for treating conditions like hypoactive sexual desire disorder (HSDD).

Quick Action: PT-141 has a rapid onset, with effects typically felt within 30–60 minutes after administration. This makes it suitable for on-demand use prior to sexual activity.

Method of Delivery and Dosage

PT-141 is typically administered via subcutaneous injection.

Dosage: The standard dosage of PT-141 is **1–2 mg per injection**, taken approximately **30–60 minutes before sexual activity**. It is recommended to start with a lower dose and adjust based on individual response and tolerance.

Cycle Duration: PT-141 can be used on an as-needed basis, typically not more than once every 24–48 hours, depending on the user's response and side effects. There is no need for continuous use, as it is designed for on-demand usage.

Kisspeptin

Kisspeptin is another peptide gaining attention for its ability to boost sexual health. It is known to stimulate the release of gonadotropin-releasing hormone (GnRH), which plays a key role in regulating reproductive hormones such as testosterone and estrogen. Kisspeptin can help improve fertility by boosting the production of these hormones.

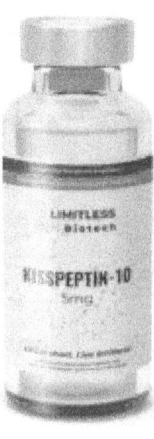

In men, it supports healthy testosterone levels, which are essential for libido and sexual performance. In women, Kisspeptin helps regulate the menstrual cycle and may improve fertility, especially in those with hormonal imbalances. By stimulating the body's natural hormonal pathways, Kisspeptin offers a more physiological approach to improving sexual health and fertility.

Benefits

Increased Testosterone Production: In men, Kisspeptin stimulates the release of GnRH, which leads to an increase in luteinizing hormone (LH) and follicle-stimulating hormone (FSH) levels. This, in turn, boosts the production of testosterone, improving libido, sexual function, and overall energy levels.

Fertility: Kisspeptin plays a key role in regulating ovulation in women, which helps to improve fertility. It helps synchronize ovulation, which is essential for conception.

Sperm Production: In men, Kisspeptin boosts the production of sperm, improving sperm count and motility.

Regulation of Reproductive Health: Kisspeptin supports the overall function of the reproductive system, making it useful for individuals with hormonal imbalances or reproductive health issues, such as polycystic ovary syndrome (PCOS) or male hypogonadism.

Method of Delivery and Dosage

Kisspeptin is typically administered via **subcutaneous** injection.

Dosage: The standard dosage of Kisspeptin for stimulating testosterone and fertility ranges from **100–200 mcg per injection**, administered **1–2 times per day**.

Cycle Duration: Kisspeptin can be used in cycles of **4–6 weeks** for testosterone and fertility improvement. It is often used as part of a fertility protocol in both men and women, with cycles suited to the individual's reproductive health needs.

Melanotan II

Melanotan II are synthetic analogs of the naturally occurring alpha-melanocyte-stimulating hormone (α-MSH), which is involved in the regulation of skin pigmentation. While **Melanotan II** were initially developed to promote tanning by increasing melanin production, it has gained additional popularity for its effects on sexual function and libido enhancement. Melanotan II works on the melanocortin system, which affects sexual desire. While its primary use is for achieving a tan, many users report increased libido as a welcome side effect. Melanotan II, has shown to increase sexual desire and erectile function in men, making it a versatile peptide for those seeking benefits in both tanning and sexual health.

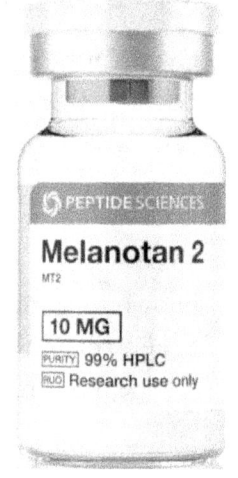

It is worth noting, however, that Melanotan II should be used carefully, as it can cause other side effects such as nausea in some users.

Benefits

Skin Tanning: Melanotan II promote the production of melanin in the skin, leading to a natural tan without excessive sun exposure. This can help protect the skin from UV damage.

Increased Libido and Sexual Arousal: Melanotan II stimulates the melanocortin receptors in the brain that are involved in sexual desire and arousal. Users often report heightened libido and improved erectile function, making it a popular choice for individuals seeking to improve sexual health.

Erectile Function: In addition to increasing libido, Melanotan II has shown to improve erectile function in men, even in those who do not respond well to traditional ED treatments. It works by increasing sexual arousal at the brain level, rather than directly affecting blood flow like PDE5 inhibitors (Viagra).

Protection Against Sunburn: By increasing melanin levels, Melanotan I and II can help protect the skin from sunburn and reduce the risk of UV-related skin damage.

Method of Delivery and Dosage

Melanotan II is administered via **subcutaneous** injection.

Dosage: The dosage for libido enhancement ranges from **0.25–1 mg per injection**, taken **every other day**.

Cycle Duration: Melanotan II is often used on a more intermittent basis, depending on the user's goals or sexual health.

5.6 Peptides for Immunity

Maintaining a strong immune system is important for overall health, especially as we age when the immune system weakens, making it harder to fight off infections and diseases. Peptides can help boost immune function by boosting the body's natural defenses, promoting faster recovery from infections, and reducing inflammation. This makes them valuable for people looking to strengthen their immune systems, particularly those with weakened immunity or autoimmune conditions.

Rather than relying solely on medications that can suppress other bodily functions, peptides help strengthen the body's own defense mechanisms, making it better equipped to fend off illnesses and recover from infections.

Thymosin Alpha-1

Thymosin Alpha-1 (Tα1) is a naturally occurring peptide derived from the thymus gland, an organ that helps in the development and regulation of the immune system. **Thymosin Alpha-1** is one of the most effective peptides for boosting immunity. It works by stimulating the production of T-cells (a type of white blood cell), which are a key component of the immune system responsible for fighting infections and protecting the body from harmful pathogens. Thymosin Alpha-1 has been used in the treatment of various conditions, including viral infections, autoimmune diseases, and even cancer. By increasing the immune system's ability to respond to threats, Thymosin Alpha-1 helps individuals recover more quickly from illness and protects against future infections.

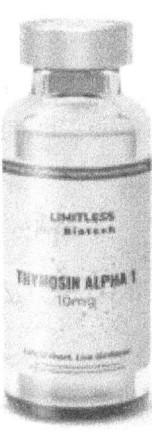

Benefits

Immune System Activation: Thymosin Alpha-1 boosts the activity of T-cells, dendritic cells, and other immune cells, increasing the body's defense mechanisms against infections, bacteria, and viruses.

Treatment for Chronic Infections: Thymosin Alpha-1 is particularly effective in treating chronic viral infections such as hepatitis B, hepatitis C, and HIV. It helps the body clear infections that are otherwise difficult to treat.

Cancer Treatment Support: By improving immune function, Thymosin Alpha-1 has been used as an adjunct therapy in cancer treatment. It helps the immune system recognize and attack cancer.

Autoimmune Disease Management: Thymosin Alpha-1 has immunomodulatory effects, meaning it can balance the immune response. This is particularly useful in autoimmune diseases, where the immune system mistakenly attacks the body's own tissues.

Vaccine Adjuvant: Thymosin Alpha-1 has been shown to improve the efficacy of vaccines by boosting the immune response, making it particularly valuable during times of widespread infections or immunization campaigns.

Method of Delivery and Dosage

Thymosin Alpha-1 is administered via **subcutaneous** injection.

Dosage: The standard dosage of Thymosin Alpha-1 is **1.5–3.2 mg per week**, divided into **2–3 injections**. In cases of chronic infection or immune deficiency, the dosage may be adjusted based on the severity of the condition.

Cycle Duration: Thymosin Alpha-1 is often used in cycles of **4–12 weeks**. In cases of chronic infection, longer cycles may be necessary, with breaks in between to assess immune function.

LL-37

Another peptide with strong immune-boosting properties is **LL-37**, an antimicrobial peptide that helps the body fight off bacterial, viral, and fungal infections. LL-37 works by disrupting the membranes of harmful pathogens, making it harder for them to survive in the body. It is known for its ability to not only kill pathogens but also modulate the immune system. This peptide is particularly useful for people with chronic infections or those who are more susceptible to illness due to a weakened immune system.

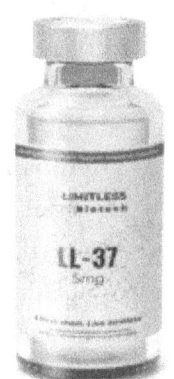

In addition to its antimicrobial properties, LL-37 also improves wound healing, reduces inflammation, making it useful for managing autoimmune and inflammatory conditions.

Benefits

Broad-Spectrum Antimicrobial Effects: LL-37 is effective against a wide variety of pathogens, including bacteria, viruses, and fungi.

Immune Modulation: In addition to its antimicrobial properties, LL-37 modulates the immune system, helping to balance immune responses and reduce excessive inflammation, which can be damaging in autoimmune diseases.

Wound Healing: LL-37 promotes tissue repair and accelerates wound healing, making it useful for individuals recovering from surgery, injuries, or chronic wounds.

Anti-Inflammatory Effects: LL-37 reduces inflammation by modulating the release of pro-inflammatory cytokines. This makes it beneficial for treating inflammatory conditions such as arthritis, psoriasis, and inflammatory bowel disease.

Protection Against Drug-Resistant Bacteria: LL-37 is effective against antibiotic-resistant bacteria, making it a valuable alternative or adjunct to traditional antibiotics in treating difficult infections.

Recommended Dosage

LL-37 is administered via **subcutaneous** injection.

Dosage: The typical dosage of LL-37 is **100 mcg per injection**, taken **2 times daily**, once in the morning and once at night.

Cycle Duration: LL-37 is commonly used in **2–4-week cycles**, depending on the severity of the infection or immune condition being treated.

VIP

Vasoactive Intestinal Peptide (VIP) is a neuropeptide that helps in regulating lung function, reducing inflammation, and modulating the immune response. **VIP** is known for its ability to relax smooth muscles, dilate blood vessels, and reduce pulmonary inflammation, making it particularly valuable for individuals with respiratory conditions such as asthma, chronic obstructive pulmonary disease (COPD), and pulmonary arterial hypertension (PAH).

VIP's anti-inflammatory properties extend beyond the lungs, as it helps reduce systemic inflammation, protect against autoimmune diseases, and support overall immune function. Its unique ability to improve lung health while regulating immune activity makes VIP a highly sought-after peptide for individuals with respiratory issues or chronic inflammation.

Benefits

Lung Health Support: VIP improves lung function by dilating airways, reducing pulmonary inflammation, and promoting healthy blood flow in the lungs. It is often used to treat conditions such as asthma, COPD, and pulmonary hypertension.

Anti-Inflammatory Effects: VIP reduces inflammation in the lungs and throughout the body by modulating cytokine production and immune cell activity. This makes it beneficial for individuals with inflammatory conditions such as arthritis, inflammatory bowel disease, and autoimmune disorders.

Immune Regulation: VIP helps balance the immune response, preventing excessive inflammation while promoting an appropriate defense against infections and pathogens. It is particularly valuable in cases of autoimmune diseases, where the immune system attacks healthy tissues.

Improved Oxygenation: By dilating blood vessels and increasing blood flow to the lungs, VIP improves oxygen delivery to the body's tissues, enhancing overall energy levels and physical performance.

Recommended Dosage

VIP is typically administered via **subcutaneous** injection, though it can also be delivered **intranasally** for respiratory benefits.

Dosage: The standard dosage of VIP is **100–500 mcg per injection**, taken **1–2 times per day**.

The recommended **Intranasal Dosage is 50 mcg** sprayed within each nostril up to **4 times per day**.

Cycle Duration: VIP can be used continuously or in cycles, depending on the user's health condition. For chronic respiratory conditions, long-term use may be necessary to maintain lung health and reduce inflammation.

KPV

KPV is a tripeptide composed of lysine, proline, and valine, known for its strong anti-inflammatory and immune-regulating properties. It has gained attention for its ability to reduce inflammation and promote healing in a variety of conditions, including inflammatory bowel disease, psoriasis, and other autoimmune disorders. KPV works by inhibiting pro-inflammatory cytokines, thereby reducing inflammation and supporting tissue repair.

KPV is often used as an adjunct treatment for inflammatory and autoimmune conditions, offering a natural approach to reducing chronic inflammation without the side effects associated with traditional anti-inflammatory medications.

Benefits

Potent Anti-Inflammatory Effects: KPV is highly effective at reducing inflammation by inhibiting the production of pro-inflammatory cytokines. This makes it valuable for treating conditions such as arthritis, psoriasis, and inflammatory bowel disease.

Immune Modulation: KPV helps regulate the immune system, preventing excessive immune responses that can lead to autoimmune flare-ups or chronic inflammation.

Wound Healing: KPV promotes tissue repair and accelerates wound healing, making it useful for individuals recovering from injuries or surgery.

Treatment for Skin Conditions: KPV has been shown to improve skin health by reducing inflammation and promoting healing in conditions such as eczema, psoriasis, and acne.

Method of Delivery and Dosage

KPV can be administered in several forms, including **subcutaneous** injections, **oral capsules**, or **topical creams**.

Dosage: The standard dosage of KPV is **1–2 mg per injection**, taken **1–2 times per day**. For inflammatory skin conditions, KPV can be applied topically in cream form, usually **once daily**.

Cycle Duration: KPV is typically used in cycles of **4–6 weeks**, depending on the user's condition and response to the peptide.

ARA-290

ARA-290 is a synthetic peptide derived from erythropoietin (EPO); a hormone involved in the production of red blood cells. However, unlike EPO, ARA-290 does not affect red blood cell production but instead focuses on promoting nerve repair, reducing inflammation, and modulating the immune system. It has been shown to improve symptoms in conditions like sarcoidosis, chronic pain, and neuropathy, making it a valuable peptide for individuals dealing with nerve damage and chronic inflammatory conditions.

ARA-290's unique ability to protect and repair nerves, reduce inflammation, and modulate immune responses makes it a promising option for treating autoimmune disorders and neuroinflammatory conditions.

Benefits

Nerve Repair and Protection: ARA-290 promotes the repair and regeneration of damaged nerves, making it useful for conditions such as neuropathy, chronic pain, and neurodegenerative diseases.

Anti-Inflammatory Effects: ARA-290 reduces inflammation by modulating the immune system and inhibiting pro-inflammatory cytokines. This makes it beneficial for treating chronic inflammatory conditions, such as sarcoidosis or autoimmune diseases.

Improved Pain Management: ARA-290 has been shown to reduce chronic pain associated with nerve damage, offering relief for individuals with neuropathic pain or other pain syndromes.

Immune System Modulation: By balancing the immune response, ARA-290 helps prevent excessive inflammation while supporting the body's ability to fight off infections and repair damaged tissues.

Recommended Dosage

ARA-290 is administered via **subcutaneous** injection, typically in the abdominal area.

Dosage: The standard dosage of ARA-290 is **5 mg per injection**, taken **once daily** for nerve repair and immune modulation. Lower doses may be used for chronic inflammation management.

Cycle Duration: ARA-290 is typically used in cycles of **4–6 weeks**, depending on the condition being treated and the user's response to the peptide.

SS-31

SS-31, also known as **Elamipretide,** is a mitochondrial-targeting peptide that has gained attention for its ability to protect and repair mitochondria, the energy-producing organelles in cells. By improving mitochondrial function, SS-31 helps reduce oxidative stress, improve cellular energy production, and support overall health and longevity. Mitochondrial dysfunction is a hallmark of aging and many chronic diseases, including neurodegenerative disorders, cardiovascular diseases, and immune deficiencies.

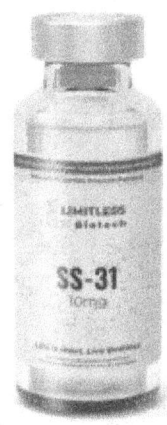

SS-31's ability to restore mitochondrial health and reduce inflammation makes it a powerful peptide for individuals seeking to boost immune function, protect against age-related diseases, and improve overall vitality.

Benefits

Improved Mitochondrial Function: SS-31 improves mitochondrial energy production, reducing oxidative stress and improving overall cellular health. This makes it useful for individuals dealing with mitochondrial dysfunction, chronic fatigue, or neurodegenerative conditions.

Anti-Aging and Longevity: By protecting mitochondria from damage, SS-31 helps delay the aging process and reduce the risk of age-related diseases such as Alzheimer's, Parkinson's, and cardiovascular disease.

Reduced Inflammation: SS-31 has potent anti-inflammatory properties, helping to reduce chronic inflammation and support immune health.

Neuroprotection: SS-31 protects neurons from oxidative damage and supports brain health, making it beneficial for individuals dealing with neurodegenerative diseases or cognitive decline.

Method of Delivery and Dosage

SS-31 is administered via **subcutaneous injection**, usually in the abdominal area.

Dosage: The standard dosage of SS-31 is **5–10 mg per injection**, taken **once daily**.

Cycle Duration: SS-31 is typically used in cycles of **4–6 weeks**, followed by a break to assess mitochondrial health and adjust dosage as needed.

5.7 Peptides for Sleep

Good sleep is essential for overall health and well-being, yet many people struggle with sleep disorders, insomnia, or poor-quality sleep. Peptides can help improve sleep quality by regulating the body's natural sleep-wake cycles, promoting relaxation, and improving deep, restorative sleep. For individuals dealing with sleep issues, peptides offer a potential solution that targets the root causes of sleep disturbances.

DSIP (Delta Sleep-Inducing Peptide)

Delta Sleep-Inducing Peptide (DSIP) is a neuropeptide known for its ability to promote restful sleep, particularly deep sleep, which is essential for recovery and tissue repair. DSIP works by regulating the body's natural sleep-wake cycle and promoting delta wave sleep, which is the deep, restorative phase of sleep. It helps reduce stress and anxiety, two of the main factors that can interfere with sleep quality.

By calming the nervous system and encouraging relaxation, DSIP helps individuals fall asleep faster and stay asleep longer, leading to more restful and restorative sleep. DSIP is particularly useful for people who have trouble reaching deep sleep or who suffer from insomnia.

Benefits

Promotes Deep Sleep: DSIP increases the body's ability to enter and maintain deep sleep, which is needed for physical recovery, memory consolidation, and overall health.

Improved Sleep Quality: Users often report more restful and uninterrupted sleep, waking up feeling more refreshed and energized.

Stress Reduction: DSIP has been shown to reduce stress and anxiety levels, helping users relax and fall asleep more easily.

Supports Recovery: Since deep sleep is essential for tissue repair and growth hormone release, DSIP can improve recovery from intense physical activity and promote overall well-being.

Method of Delivery and Dosage

DSIP is typically administered via **subcutaneous** injection, usually before bedtime to align with the body's natural sleep cycle.

Dosage: The standard dosage of DSIP is **100–300 mcg per injection**, taken **30–60 minutes before bedtime**. For individuals with more severe sleep disturbances, higher doses may be used under medical supervision.

Cycle Duration: DSIP can be used intermittently or in cycles of **4–6 weeks**, depending on the user's needs and response to the peptide.

Epitalon

Epitalon, also known as Epithalon, is a synthetic peptide derived from the naturally occurring peptide epithalamin, which is produced in the pineal gland. Epitalon is best known for its anti-aging effects. However, it also plays a key role in regulating melatonin production, which helps in improving sleep quality.

By normalizing circadian rhythms and promoting the natural release of melatonin, Epitalon helps users achieve more restful and rejuvenating sleep, especially in older adults who often experience declining melatonin levels.

Benefits

Improved Sleep Quality: Epitalon enhances the body's ability to produce melatonin, which regulates the sleep-wake cycle and promotes deep, restful sleep.

Regulation of Circadian Rhythms: Epitalon helps normalize circadian rhythms, particularly in older adults who experience disrupted sleep patterns due to reduced melatonin production.

Enhanced Recovery: By promoting deeper sleep, Epitalon improves recovery from physical exertion and supports overall health.

Method of Delivery and Dosage

Epitalon is administered via subcutaneous injection, typically before bedtime to enhance melatonin production and improve sleep quality.

Dosage: The standard dosage of Epitalon is **1–3 mg per day**, administered for **10–20 days**. This cycle can be repeated every **6–12 months** for long-term sleep.

Cycle Duration: Epitalon is usually used in short cycles of **10–20 days**, followed by a break.

Thymosin Beta-4

Thymosin Beta-4 (TB-4), primarily known for its tissue repair and healing properties, has been found to improve sleep indirectly by speeding recovery and reducing inflammation. When the body is in a state of healing or inflammation, it can disrupt sleep patterns. TB-4's ability to reduce inflammation and promote tissue repair can help individuals experience better sleep, particularly those recovering from injuries or dealing with chronic inflammation.

Benefits

Reduced Inflammation: TB-4's anti-inflammatory effects help alleviate pain and discomfort that can disrupt sleep, particularly in individuals with chronic conditions such as arthritis or injuries.

Improved Sleep Quality: Users often report more restful sleep due to reduced pain and faster recovery from injuries, allowing the body to enter deeper stages of sleep.

Muscle Relaxation: TB-4 promotes muscle relaxation, reducing tension and promoting a more restful sleep.

Recovery: By promoting tissue repair and reducing muscle soreness, TB-4 improves recovery from physical exertion, allowing for better sleep and reduced nighttime discomfort.

Method of Delivery and Dosage

TB-4 is administered via subcutaneous injection, typically in the abdominal area or near the site of injury for localized benefits.

Dosage: The standard dosage of TB-4 for sleep and recovery is **2–5 mg per week**, divided into **2–3 injections**.

Cycle Duration: TB-4 is commonly used in **4–8-week** cycles followed by a break.

5.8 Peptides for Skin, Hair, and Aesthetics

Many peptides are used for their ability to improve the appearance of the skin, hair, and overall aesthetics. These peptides promote collagen production, reduce inflammation, and increase tissue repair, leading to healthier skin, thicker hair, and a more youthful appearance.

GHK-Cu

GHK-Cu is one of the most well-known peptides for improving skin health. It is a copper peptide that promotes collagen production, reduces wrinkles, and improves skin elasticity. Discovered in the 1970s, GHK-Cu has since become a popular ingredient in anti-aging and skincare products due to its ability to promote youthful skin, reduce fine lines, and increase hair growth. GHK-Cu also has anti-inflammatory properties, which help reduce skin redness and irritation.

This peptide is often used in anti-aging skincare products, but it can also be applied directly to wounds or scars to promote healing and reduce scarring. Additionally, GHK-Cu has been shown to improve hair growth by stimulating hair follicles and promoting a healthier scalp.

Benefits

Skin Repair and Wound Healing: GHK-Cu accelerates wound healing by promoting the regeneration of skin cells and reducing inflammation. This makes it highly effective for treating scars, cuts, and abrasions.

Collagen Production: One of GHK-Cu's most notable benefits is its ability to stimulate collagen production. Increased collagen levels help improve skin elasticity, reduce wrinkles, and restore a more youthful appearance.

Hair Growth: GHK-Cu promotes hair follicle health, encouraging new hair growth and reducing hair loss. It has been shown to improve hair thickness and density over time.

Anti-Inflammatory and Antioxidant Properties: GHK-Cu helps reduce inflammation and oxidative stress in the skin, which can lead to clearer, more radiant skin. It is particularly beneficial for individuals dealing with skin conditions like acne, eczema, or rosacea.

Method of Delivery and Dosage

GHK-Cu can be applied **topically** as part of a skincare regimen or administered via **subcutaneous** injection for systemic benefits.

Topical Dosage: GHK-Cu is typically used in serums or creams at concentrations of **0.5–1%**, applied to the skin **once or twice daily**.

Injectable Dosage: For systemic skin and hair benefits, GHK-Cu can be administered subcutaneously at a dose of **2–5 mg per injection**, usually taken **once daily** over a **4–6-week cycle**.

Argireline

Argireline is a peptide often referred to as "Botox in a bottle" due to its ability to reduce wrinkle formation. Argireline works by inhibiting muscle contractions, which reduces the appearance of fine lines and wrinkles, particularly around the eyes and forehead. Unlike Botox, Argireline can be applied topically and does not require injections, making it a convenient option for those looking for non-invasive anti-aging solutions. Argireline is commonly found in serums and creams and can be combined with other peptides to improve anti-aging effects.

Benefits:

- **Reduced Wrinkle Depth**: Inhibits neurotransmitter release to smooth fine lines and wrinkles, particularly in high-expression areas like the forehead and around the eyes.
- **Skin Firmness and Smoothness**: Improves skin texture by relaxing underlying muscles, resulting in a firmer, smoother appearance.
- **Non-Invasive Botox Alternative**: Provides Botox-like effects without injections, making it accessible for daily use in skincare.

Recommended Dosage:

- **Topical Application**: Typically formulated in concentrations of 5–10% in serums or creams for direct application to wrinkle-prone areas.

Cycle: Argireline can be applied daily as part of a skincare routine, with effects typically noticeable within a few weeks of consistent use.

PTD-DBM

PTD-DBM is a cosmetic peptide specifically targeted toward hair growth and follicle health. It works by inhibiting the protein CXXC5, which can interfere with Wnt/β-catenin signaling, a pathway essential for hair growth. By blocking this protein, PTD-DBM encourages hair follicle regeneration and improves scalp health, making it a promising treatment for hair loss and thinning. PTD-DBM is often used in conjunction with other hair-promoting treatments.

Benefits:

- **Promotes Hair Growth**: Stimulates dormant hair follicles, leading to thicker and fuller hair.
- **Improved Scalp Health**: Enhances scalp condition by supporting hair follicle health and reducing inflammation.
- **Supports Hair Follicle Regeneration**: PTD-DBM encourages the growth of new hair in thinning or bald areas by targeting specific proteins that inhibit hair follicle development.

Recommended Dosage:

- **Topical Solution** applied to the scalp at a concentration of 0.1–0.5.

- When used in clinical settings, PTD-DBM can be administered at 5–10 mg per week, depending on the extent of hair loss, through **subcutaneous injections** in the scalp.

Cycle: 8–12 weeks of consistent application, with results often visible within this period. PTD-DBM can be used in repeated cycles for continuous support in hair growth and maintenance.

BPC-157

BPC-157, while primarily known for its healing properties, can also improve skin health by promoting tissue repair and reducing inflammation. It has been used to treat wounds, burns, and scars, helping the skin heal faster and reducing the appearance of scarring. BPC-157 improves blood flow and promote tissue regeneration which helps overall skin quality and reducing the signs of aging.

BPC-157's ability to promote angiogenesis (the formation of new blood vessels) further adds to its benefits for skin repair and overall skin health.

Benefits

Wound Healing: BPC-157 speeds up the healing of skin wounds by promoting tissue regeneration and reducing inflammation. It is particularly beneficial for post-surgical recovery and the healing of burns, cuts, and abrasions.

Scar Reduction: BPC-157 helps minimize scar formation by promoting more efficient collagen synthesis and reducing fibrosis (excess tissue buildup).

Anti-Inflammatory Effects: It reduces inflammation in the skin, which can be beneficial for treating conditions such as acne, dermatitis, and other inflammatory skin disorders.

Skin Regeneration: BPC-157 supports the regeneration of skin cells, leading to smoother, healthier-looking skin over time.

Method of Delivery and Dosage

BPC-157 can be applied **topically** or administered via **subcutaneous injection**, depending on the desired effect.

Topical Dosage: When applied topically, BPC-157 is typically used in concentrations of **250–500 mcg** per application, applied **once or twice daily** to the affected area.

Injectable Dosage: For systemic wound healing and skin regeneration, the standard injectable dose of BPC-157 is **200–400 mcg per injection**, taken **once or twice daily**. Treatment cycles typically last **4–6 weeks**.

Melanotan I & II

Melanotan I and II are synthetic analogs of alpha-melanocyte-stimulating hormone (α-MSH), which regulates skin pigmentation. They are primarily used to stimulate tanning by increasing the production of melanin in the skin. Melanin is the pigment responsible for skin color, and by promoting its production, Melanotan peptides can give users a natural-looking tan without excessive sun exposure.

In addition to their tanning effects, some users report that Melanotan peptides improve skin texture and reduce the appearance of blemishes or uneven skin tone.

Benefits

Skin Tanning: Melanotan I and II stimulate melanin production, leading to a gradual and even tan with minimal sun exposure. This is especially beneficial for individuals with fair skin who are prone to burning.

UV Protection: By increasing melanin levels, Melanotan peptides provide a natural defense against UV radiation, reducing the risk of sunburn and skin damage.

Pigmentation Disorder Treatment: Melanotan I and II can help treat pigmentation disorders such as vitiligo, where areas of the skin lose pigment and become lighter.

Libido Enhancement (Melanotan II): In addition to its tanning effects, Melanotan II has been shown to enhance libido and sexual arousal by acting on melanocortin receptors in the brain.

Method of Delivery and Dosage

Melanotan peptides are administered via **subcutaneous injection,** typically in the abdominal area.

Dosage (Melanotan I): For tanning, the typical dosage of Melanotan I is **0.5–1 mg per injection**, taken **1–2 times per week**. More frequent dosing may be needed initially to build up melanin levels.

Dosage (Melanotan II): The standard dosage of Melanotan II is **0.25–1 mg per injection**, taken **every other day**.

5.9 Peptides for Women

Hormonal imbalances can affect women at different stages of life, from menstrual irregularities to menopause. Peptides offer a targeted approach to addressing these imbalances, helping women improve their overall wellness, manage symptoms, and improve their quality of life.

Kisspeptin

Kisspeptin is a peptide that plays a key role in regulating reproductive hormones, particularly by stimulating the release of gonadotropin-releasing hormone (GnRH), which in turn regulates the production of estrogen and progesterone. For women dealing with fertility issues or hormonal imbalances, Kisspeptin can help restore normal hormone levels and improve reproductive health. It has shown promise in treating conditions such as polycystic ovary syndrome (PCOS), a common cause of infertility in women.

Benefits

Fertility Enhancement: Kisspeptin stimulates ovulation by promoting the release of GnRH, LH, and FSH, improving fertility in women who struggle with ovulatory disorders.

Hormonal Balance: By regulating the release of sex hormones, Kisspeptin helps balance estrogen and progesterone levels, promoting regular menstrual cycles and reducing symptoms of hormonal imbalance.

Support for PCOS: Kisspeptin has shown promise in regulating ovulation and reducing hormonal imbalances in women with PCOS, a common cause of infertility.

Improved Sexual Health: Kisspeptin can enhance libido and sexual health by promoting healthy hormone levels and improving overall reproductive function.

Method of Delivery and Dosage

Kisspeptin is administered via **subcutaneous** injection.

Dosage: The typical dosage for Kisspeptin is **100–200 mcg per injection**, taken **1–2 times per day**.

Cycle Duration: Kisspeptin is often used in cycles of **4–6 weeks**, particularly for women trying to conceive or regulate their menstrual cycles.

Peptides for Menopause

Menopause is a natural biological process that marks the end of a woman's reproductive years, typically occurring between the ages of 45 and 55. It is characterized by a decline in estrogen and progesterone levels, leading to symptoms such as hot flashes, night sweats, mood swings, and sleep disturbances. Peptides such as **CJC-1295**, **Ipamorelin**, and **GHK-Cu** have shown promise in managing the symptoms of menopause by supporting hormonal balance, improving skin and hair health, and enhancing overall well-being.

These peptides stimulate the release of growth hormone, which can help to alleviate the effects of hormonal decline, promote better sleep, and support anti-aging processes, especially for women going through menopause.

Benefits

Symptom Relief: Peptides such as CJC-1295 and Ipamorelin help alleviate common menopause symptoms, including hot flashes, night sweats, and mood swings, by promoting hormonal balance.

Improved Skin and Hair Health: GHK-Cu supports collagen production, which can help improve skin elasticity, reduce wrinkles, and promote hair growth, addressing the aesthetic concerns often associated with menopause.

Sleep and Energy Levels: By improving growth hormone release and regulating sleep cycles, these peptides help women achieve better sleep quality, increased energy, and improved overall well-being.

Method of Delivery and Dosage

Peptides for menopause are typically administered via subcutaneous injection.

Dosage: CJC-1295 and Ipamorelin are typically dosed at **100–300 mcg per injection**, taken **once daily**, while GHK-Cu is dosed at **2–5 mg per injection**, usually taken **once daily** for skin and hair benefits.

Cycle Duration: These peptides are often used in cycles of **8–12 weeks**.

PT-141

PT-141 (Bremelanotide) is a powerful peptide that increase sexual desire and arousal in both men and women by acting on melanocortin receptors in the brain. For women, PT-141 offers an effective treatment for low libido, hypoactive sexual desire disorder (HSDD), and sexual dysfunction, especially those related to hormonal changes, such as menopause. Unlike traditional libido treatments that focus solely on physical performance, PT-141 targets the brain's arousal pathways to increase sexual desire.

Benefits

- **Increased Libido**: PT-141 directly stimulates sexual desire and arousal, making it particularly effective for women with low libido or hypoactive sexual desire disorder (HSDD).

- **Improved Sexual Satisfaction**: By improving sexual response, PT-141 can improve overall sexual satisfaction, making it easier for women to achieve orgasm and enjoy a more fulfilling sex life.

- **Quick Action**: PT-141 has a quick onset of action, typically within 30–60 minutes, making it suitable for use prior to sexual activity.

Method of Delivery and Dosage

PT-141 is administered via **subcutaneous** injection, typically before sexual activity.

- **Dosage**: The standard dosage of PT-141 for libido enhancement is **1–2 mg per injection**, taken approximately **30–60 minutes before sexual activity**.

- **Cycle Duration**: PT-141 can be used on an as-needed basis, typically not more than once every 24–48 hours.

5.10 Peptides for Men

As men age, they experience a decline in hormone levels, particularly testosterone. This condition, often referred to as andropause or male menopause, can lead to symptoms such as low energy, decreased libido, mood swings, and a reduction in muscle mass. Peptides are increasingly being used to help men address these hormonal imbalances and maintain their health and vitality as they age.

Gonadorelin

Gonadorelin is a peptide that stimulates the production of luteinizing hormone (LH), which is responsible for regulating testosterone production in men. By increasing LH levels, Gonadorelin helps stimulate the body's natural production of testosterone, making it an effective treatment for men dealing with low testosterone levels. It can be used as an alternative to traditional testosterone replacement therapy (TRT) for men who want to restore their natural testosterone levels without relying on synthetic hormones.

Benefits

Increased Testosterone Production: Gonadorelin stimulates the release of LH and FSH, leading to a natural increase in testosterone levels. This helps improve libido, energy, and muscle mass.

Fertility: In addition to boosting testosterone, Gonadorelin supports sperm production, improving fertility in men with low sperm counts or poor sperm motility.

Mood and Mental Clarity: By restoring hormonal balance, Gonadorelin can help improve mood, reduce symptoms of depression, and enhance cognitive function.

Method of Delivery and Dosage

Gonadorelin is administered via subcutaneous or intramuscular injection.

Dosage: The typical dosage of Gonadorelin for testosterone stimulation is **100–200 mcg per injection**, taken **1–2 times per day**.

Cycle Duration: Gonadorelin is often used in cycles of **4–6 weeks**.

Kisspeptin

Kisspeptin also plays a role in male hormonal health by regulating the release of GnRH, which in turn stimulates testosterone production. Kisspeptin has been shown to improve fertility in men by promoting healthy sperm production and improving overall reproductive health.

For men experiencing a decline in testosterone levels due to aging or other factors, Kisspeptin can help restore balance and improve libido, sexual performance, and mood.

Benefits

Increased Testosterone Levels: Kisspeptin boosts the production of testosterone, improving libido, energy, and sexual performance.

Fertility: By stimulating sperm production, Kisspeptin can improve fertility in men with low sperm counts or poor sperm motility.

Improved Sexual Health: Kisspeptin enhances sexual desire and performance, making it a valuable tool for men with low libido or erectile dysfunction.

Method of Delivery and Dosage

Kisspeptin is administered via **subcutaneous** injection.

Dosage: The typical dosage of Kisspeptin for fertility and testosterone enhancement is **100–200 mcg per injection**, taken **1–2 times per day**.

Cycle Duration: Kisspeptin is commonly used in **4–6-week cycles**.

PT-141

PT-141 is another peptide that has proven beneficial for men dealing with erectile dysfunction or low libido. Unlike traditional erectile dysfunction medications, which focus on improving blood flow, PT-141 works by stimulating sexual desire. Unlike traditional ED treatments that focus solely on increasing blood flow to the penis, PT-141 works by stimulating sexual desire and increasing the body's natural arousal mechanisms.

It has proven to be effective for men with erectile dysfunction (ED), particularly those who have not responded well to PDE5 inhibitors (like Viagra).

Benefits

Improved Erectile Function: PT-141 improves erectile function by increasing sexual arousal, making it particularly effective for men with ED caused by psychological or hormonal factors.

Increased Libido: PT-141 boosts sexual desire, making it easier for men to achieve and maintain erections during sexual activity.

Rapid Onset: PT-141 has a quick onset of action, typically within 30–60 minutes, making it suitable for on-demand use before sexual activity.

Method of Delivery and Dosage

PT-141 is administered via **subcutaneous** injection, typically before sexual activity.

Dosage: The standard dosage of PT-141 is **1–2 mg per injection**, taken **30–60 minutes before sexual activity**.

Cycle Duration: PT-141 is used on an as-needed basis and should not be taken more than once every 24–48 hours.

CHAPTER 6. PEPTIDE STACKS AND COMBINATIONS

Combining peptides, known as peptide stacking, is a popular strategy used by individuals looking to improve the effectiveness of their peptide therapy. Peptide stacking allows users to achieve more significant results and benefits than using a single peptide alone, whether their goal is muscle growth, fat loss, anti-aging, cognitive enhancement, or immune support. Stacks typically involve two or more peptides that are cycled together for a specific period, followed by a break or a "rest cycle" to allow the body to reset. These cycles can vary depending on the user's goals and the peptides being stacked.

When done correctly, peptide stacking allows users to address multiple physiological processes simultaneously, leading to synergistic effects that exceed the benefits of using a single peptide. However, to achieve the best results, it's important to understand how different peptides interact with one another and how to cycle them effectively to avoid diminishing returns or side effects.

When stacking peptides, the goal is to combine peptides that work via different but complementary pathways to achieve a broader range of effects. This allows for greater overall results in areas like muscle growth, fat loss, and recovery. For example, stacking growth hormone-releasing peptides (GHRPs) with peptides that promote tissue repair can result in better recovery from workouts and more significant muscle growth.

6.1 Peptides Stacks/Combos for Fat Loss

Ipamorelin + CJC-1295

Combining Ipamorelin with CJC-1295 creates a powerful fat-loss stack. Ipamorelin stimulates growth hormone release, while CJC-1295 increases the duration of this release. Together, they boost metabolism and aid in fat reduction, especially when paired with diet and exercise.

Benefits:

- **Fat Breakdown:** Stimulates lipolysis by releasing growth hormone.
- **Muscle Preservation:** Both peptides help in retaining muscle mass during fat loss.
- **Increased Energy and Metabolism:** Users experience a rise in metabolic rate, burning more calories even at rest.

Recommended Dosage:

- **Ipamorelin:** 200–300 mcg per injection, taken 1–2 times daily.
- **CJC-1295:** 1 mg per injection, twice a week.

Ipamorelin + CJC-1295 + AOD-9604

This combination leverages the fat-burning properties of growth hormone (GH) stimulation with the targeted fat loss effects of AOD-9604. Ipamorelin and CJC-1295 both trigger growth hormone release, aiding in fat metabolism and muscle retention. AOD-9604 increases the fat-burning process without raising blood sugar levels, making it ideal for those aiming to lose weight while preserving lean muscle mass.

Benefits:

- **Fat Breakdown**: Ipamorelin and CJC-1295 promote lipolysis through GH stimulation. AOD-9604 adds an extra layer of fat reduction, particularly around stubborn areas like the abdomen.
- **Muscle Preservation**: While focusing on fat reduction, the stack helps maintain lean muscle mass.
- **Increased Metabolism**: Growth hormone's effects on metabolism allow for calorie burning even at rest, while AOD-9604 provides specific fat-targeting mechanisms.

Dosage:

- **Ipamorelin**: 200–300 mcg per injection, taken 1–2 times daily.
- **CJC-1295**: 1 mg per injection, twice weekly.
- **AOD-9604**: 300 mcg daily, via subcutaneous injection.

Semaglutide + MOTS-C + Tesamorelin

This stack combines **Semaglutide**, a GLP-1 receptor agonist that reduces appetite and promotes weight loss, **MOTS-C**, a mitochondrial peptide that improves fat oxidation, and **Tesamorelin**, which specifically targets visceral fat. Together, these peptides create a potent fat-loss stack for individuals looking to reduce fat and manage metabolic health.

Benefits:

- **Appetite Control**: Semaglutide helps reduce cravings and calorie intake by delaying gastric emptying.
- **Fat Oxidation**: MOTS-C boosts mitochondrial function, allowing for more efficient fat burning during exercise.
- **Visceral Fat Reduction**: Tesamorelin is particularly effective in reducing belly fat, improving body composition.

Method of Delivery and Dosage:

- **Semaglutide**: 0.25–1.0 mg weekly, via subcutaneous injection.
- **MOTS-C**: 10 mg per week, divided into 2–3 injections.
- **Tesamorelin**: 2 mg daily, via subcutaneous injection.

Cycle: 12–16 weeks, with periodic breaks to monitor insulin sensitivity and metabolic responses.

Tirzepatide + Tesofensine + 5-Amino 1MQ

Tirzepatide combines GLP-1 and GIP receptor stimulation to promote fat loss and metabolic control. Paired with **Tesofensine**, which suppresses appetite, and **5-Amino 1MQ**, which aids cellular metabolism, this stack offers powerful fat-burning potential while maintaining energy and focus during a weight-loss regimen.

Benefits:

- **Appetite and Metabolism**: Tirzepatide and Tesofensine work together to reduce hunger while boosting the body's fat-burning capacity.
- **Fat Metabolism**: 5-Amino 1MQ stimulates fat oxidation by targeting cellular pathways involved in metabolism.
- **Sustained Weight Loss**: This stack ensures steady fat loss with preserved lean muscle mass.

Dosage:

- **Tirzepatide**: 2.5–15 mg weekly, via subcutaneous injection.
- **Tesofensine**: 0.5 mg orally, daily.
- **5-Amino 1MQ**: 50–100 mg orally, per day.

Cycle: 8–12 weeks for best results, with breaks to reset metabolic responses.

Tesamorelin + CJC-1295 + MK-677

This stack is ideal for individuals looking to burn fat while also gaining lean muscle mass. **Tesamorelin** and **CJC-1295** both stimulate growth hormone release, promoting fat loss, while **MK-677** increases appetite and supports muscle growth, making this a balanced stack for body recomposition.

Benefits:

- **Fat Reduction and Muscle Gain**: Tesamorelin and CJC-1295 trigger fat metabolism while maintaining or increasing muscle mass.
- **Increased Appetite and Recovery**: MK-677 improves appetite and supports recovery from intense workouts.
- **Improved Metabolism**: The combination accelerates metabolism, ensuring efficient fat burning throughout the day.

Method of Delivery and Dosage:

- **Tesamorelin**: 2 mg daily.
- **CJC-1295**: 1 mg twice weekly.

- **MK-677**: 10–25 mg daily, orally.

Cycle: 12–16 weeks with a break.

AOD-9604 + Ipamorelin + Tirzepatide

This stack harnesses the fat-burning capabilities of **AOD-9604** while **Ipamorelin** and **Tirzepatide** further accelerate fat loss and enhance metabolism. It's an ideal stack for individuals who need strong appetite control and targeted fat reduction.

Benefits:

- **Targeted Fat Reduction**: AOD-9604 focuses on stubborn fat areas like the abdomen.
- **Appetite Suppression**: Tirzepatide curbs cravings, helping users adhere to calorie-restricted diets.
- **Fat Metabolism**: Ipamorelin increases the breakdown of fat stores, enhancing overall body composition.

Dosage:

- **AOD-9604**: 300 mcg daily, via injection.
- **Ipamorelin**: 200–300 mcg, 1–2 times daily, via injection.
- **Tirzepatide**: 5 mg weekly, via injection.

NB: This list is not exhaustive; they can be adjusted based on your personal needs.

6.2 Peptides Stacks/Combos for Muscle Growth

CJC-1295 + Ipamorelin + IGF-1 LR3

This powerful stack/combination targets growth hormone production, muscle cell proliferation, and recovery. **CJC-1295** provides sustained growth hormone release, **Ipamorelin** triggers immediate growth hormone spikes, and **IGF-1 LR3** promotes muscle cell growth and regeneration. Together, they form a strong combination for individuals aiming to increase muscle mass, improve strength, and accelerate recovery.

Benefits:

- **Growth Hormone Release:** CJC-1295 maintains elevated growth hormone levels, supporting long-term muscle growth.
- **Muscle Repair and Growth:** IGF-1 LR3 increases muscle cell proliferation, speeding up repair after workouts and promoting denser muscle growth.
- **Fat Reduction:** Growth hormone's effects on metabolism help burn fat while preserving muscle mass.

Dosage:

- **CJC-1295:** 1000 mcg twice weekly, via subcutaneous injection.
- **Ipamorelin:** 200–300 mcg, 1–2 times daily, via subcutaneous injection.
- **IGF-1 LR3:** 20–50 mcg daily, via subcutaneous injection, preferably post-workout.

Cycle: 8–12 weeks with breaks of 4–6 weeks.

CJC-1295 + Ipamorelin + BPC-157

This stack combines **CJC-1295** and **Ipamorelin** for sustained and immediate growth hormone release, paired with **BPC-157** to promote rapid tissue repair and reduce inflammation. **CJC-1295** provides a steady release of growth hormone, while **Ipamorelin** stimulates a quick burst, enhancing muscle growth and recovery. **BPC-157** complements these by aiding in healing and repair, making this combination ideal for athletes and bodybuilders focused on strength and recovery.

Benefits:

- **Muscle Growth**: CJC-1295 and Ipamorelin stimulate growth hormone release, aiding in muscle development and maintenance.
- **Recovery**: BPC-157 speeds up tissue repair, reducing inflammation and supporting faster recovery after workouts.
- **Reduced Injury Risk**: BPC-157 supports joint, ligament, and tendon health, making it ideal for preventing overuse injuries.

Dosage:

- **CJC-1295**: 1000 mcg twice weekly, via subcutaneous injection.
- **Ipamorelin:** 200–300 mcg, 1–2 times daily, via subcutaneous injection.
- **BPC-157**: 200–500 mcg daily, via subcutaneous injection.

Cycle: 8–12 weeks, with a break of 4–6 weeks.

CJC-1295 + GHRP-2 + BPC-157

This muscle growth and recovery stack combines **CJC-1295** and **GHRP-2** to stimulate growth hormone release while **BPC-157** promotes tissue repair. **CJC-1295** provides a long-acting growth hormone boost, and **GHRP-2** offers immediate GH spikes, which improves muscle growth and improves recovery speed. **BPC-157** helps reduce inflammation and supports joint health, which is particularly useful during intense training.

Benefits:

- **Muscle Mass and Fat Loss**: CJC-1295 and GHRP-2 stimulate growth hormone, promoting lean muscle growth and helping reduce body fat.
- **Accelerated Recovery**: BPC-157 helps repair damaged tissues and reduce inflammation, aiding in faster recovery.
- **Enhanced Joint Health**: BPC-157 supports ligaments and tendons, reducing the risk of injury during heavy lifting or intense exercise.

Dosage:

- **CJC-1295:** 1000 mcg twice weekly, via subcutaneous injection.
- **GHRP-2:** 100–300 mcg, 1–2 times daily, via subcutaneous injection.
- **BPC-157:** 200–500 mcg daily, via subcutaneous injection.

Cycle: 8–12 weeks, with breaks in between.

CJC-1295 + GHRP-6 + BPC-157

This combo/stack combines **CJC-1295** and **GHRP-6** to promote growth hormone release with **BPC-157** for tissue healing and inflammation reduction. **CJC-1295** provides long-lasting GH release, while **GHRP-6** induces a strong appetite, supporting muscle gains for those aiming to bulk up. **BPC-157** aids in tissue repair, making this stack beneficial for muscle growth, recovery, and injury prevention.

Benefits:

- **Growth Hormone Release and Muscle Development**: CJC-1295 and GHRP-6 work together to support muscle growth, reduce fat, and improve recovery.
- **Improved Appetite for Bulking**: GHRP-6 stimulates appetite, making it easier to meet increased caloric needs for muscle growth.
- **Faster Healing and Reduced Inflammation**: BPC-157 supports recovery of muscles, tendons, and ligaments, reducing downtime between training sessions.

Dosage:

- **CJC-1295:** 1000 mcg twice weekly, via subcutaneous injection.
- **GHRP-6:** 100–300 mcg, 1–2 times daily, via subcutaneous injection.
- **BPC-157:** 200–500 mcg daily, via subcutaneous injection.

Cycle: 8–12 weeks, with a 4-week break in between cycles.

MK-677 + GHRP-6 + PEG-MGF

This combo/stack combines **MK-677**, an oral growth hormone secretagogue, with **GHRP-6**, a potent GHRP that increases GH secretion, and **PEG-MGF**, which boosts muscle repair. This stack is designed for individuals focused on bulking, as it promotes both muscle gain and improved recovery.

Benefits:

- **Lean Muscle Gain:** MK-677 and GHRP-6 stimulate GH release, supporting muscle hypertrophy and retention.
- **Improved Recovery:** PEG-MGF aids in muscle repair by increasing satellite cell activation, accelerating the recovery process after intense training.
- **Appetite:** GHRP-6 boosts appetite, supporting increased caloric intake necessary for muscle growth.

Dosage:

- **MK-677 (Oral):** 10–25 mg daily.
- **GHRP-6:** 100–200 mcg, 1–2 times daily, via subcutaneous injection.
- **PEG-MGF:** 200–400 mcg, 2–3 times weekly, via subcutaneous injection.

Cycle: 12–16 weeks for best results, followed by a 4-week break to reset growth hormone (GH) receptors.

TB-500 + BPC-157 + CJC-1295

This stack/combination is focused on muscle recovery and repair, making it useful for athletes or bodybuilders recovering from injuries or those undergoing high-intensity training. **TB-500** and **BPC-157** speed up tissue repair, while **CJC-1295** boosts growth hormone to further aid recovery and muscle growth.

Benefits:

- **Injury Recovery:** TB-500 and BPC-157 accelerate the healing of muscle, tendon, and ligament injuries.
- **Tissue Repair:** CJC-1295 supports long-term muscle regeneration by increasing growth hormone levels.
- **Improved Muscle Endurance:** This stack helps muscles recover faster, allowing for more

Dosage:

- **TB-500:** 2–5 mg weekly, via subcutaneous injection.
- **BPC-157:** 200–500 mcg, 1–2 times daily, via subcutaneous injection.
- **CJC-1295:** 1000 mcg twice weekly, via subcutaneous injection.

IGF-1 DES + Follistatin-344 + GHRP-2

This potent muscle-building stack/combo focuses on muscle cell growth and inhibiting myostatin, a protein that limits muscle development. **IGF-1 DES** and **Follistatin-344** promote muscle hypertrophy by encouraging new muscle fiber growth and blocking myostatin. **GHRP-2** supports growth hormone secretion to further aid in muscle repair and growth.

Benefits:

- **Muscle Hypertrophy:** IGF-1 DES and Follistatin-344 significantly increase muscle cell growth, leading to rapid size and strength gains.
- **Inhibition of Myostatin:** Follistatin-344 blocks myostatin, allowing for unrestrained muscle growth.
- **Increased Growth Hormone:** GHRP-2 triggers natural GH release, increasing muscle repair and performance.

Dosage:

- **IGF-1 DES:** 50–100 mcg daily, via **subcutaneous** or intramuscular injection.
- **Follistatin-344:** 100 mcg daily for 10 days, via **subcutaneous** or intramuscular injection.
- **GHRP-2:** 100–200 mcg, 1–2 times daily, via subcutaneous injection.

Cycle: 8–10 weeks for optimal muscle gain, followed by a break of 4–6 weeks.

Hexarelin + Ipamorelin + IGF-1 LR3

Combining **Hexarelin**, one of the most potent growth hormone-releasing peptides, with **Ipamorelin** and **IGF-1 LR3**, this stack increases both short- and long-term growth hormone release. **Hexarelin** and **Ipamorelin** together ensure a powerful GH spike, while **IGF-1 LR3** promotes muscle growth and repair, making this stack effective for muscle building and body recomposition.

Benefits:

- **Powerful GH Release:** Hexarelin provides a strong growth hormone surge, complemented by Ipamorelin's gradual and sustained release.
- **Muscle Growth:** IGF-1 LR3 promotes new muscle cell growth and helps repair micro-tears caused by intense training.
- **Improved Body Composition:** This stack supports muscle hypertrophy while reducing body fat.

Dosage:

- **Hexarelin:** 100–200 mcg, 1–2 times daily, via subcutaneous injection.
- **Ipamorelin:** 200–300 mcg, 1–2 times daily, via subcutaneous injection.

- **IGF-1 LR3:** 20–50 mcg daily, via subcutaneous injection.

Cycle: 8–12 weeks with a 4-week break.

Hexarelin + TB-500 + PEG-MGF

This combo/stack is designed for significant muscle growth and recovery. **Hexarelin** is a powerful growth hormone-releasing peptide, **TB-500** supports tissue repair and reduces inflammation, and **PEG-MGF** (Pegylated Mechano Growth Factor) stimulates muscle cell repair and growth. This stack is ideal for athletes and bodybuilders who aim to optimize muscle gains, improve recovery speed, and prevent injuries.

Benefits:

- **Maximized Growth Hormone Release**: Hexarelin provides a potent GH boost, promoting muscle development and reducing fat stores.
- **Tissue and Muscle Repair**: TB-500 accelerates healing and supports connective tissue health, making it excellent for injury prevention.
- **Increased Muscle Cell Growth**: PEG-MGF promotes the growth of muscle fibers and aids in recovery after strenuous exercise.

Method of Delivery and Dosage:

- **Hexarelin**: 100–200 mcg, 1–2 times daily, via subcutaneous injection.
- **TB-500**: 2–5 mg weekly, via subcutaneous injection.
- **PEG-MGF**: 200–400 mcg, 2–3 times weekly, injected directly into the muscle post-workout.

Cycle: 8–12 weeks, with a 4-week break to allow growth hormone receptors to reset.

6.3 Brain Health and Cognitive Performance Stacks/Combos

Semax + Selank + Cerebrolysin

This combination of **Semax**, **Selank**, and **Cerebrolysin** focuses on improving cognitive function, memory retention, and neuroprotection. **Semax** is a nootropic peptide known for improving focus and cognitive performance, while **Selank** helps reduce anxiety and improves mood. **Cerebrolysin**, a neuropeptide blend, protects brain cells and promotes brain repair, making this stack ideal for boosting mental clarity and long-term brain health.

Benefits:

- **Improved Focus and Memory**: Semax improves focus, attention, and learning capacity. It is often used by individuals seeking sharper mental performance.
- **Reduced Anxiety and Stress**: Selank works as an anxiolytic, helping reduce stress and anxiety, leading to better overall cognitive function.

- **Neuroprotection and Brain Repair**: Cerebrolysin supports brain cell repair and protects neurons from damage, making it beneficial for both cognitive enhancement and neuroprotection.

Recommended Dosage:

- **Semax:** 300 mcg, 2–3 times daily, via nasal spray or injection. **Nasal spray** is the most common method.
- **Selank :** 200–300 mcg, 2–3 times daily, via nasal spray or injection.
- **Cerebrolysin (Injection):** 5–10 ml, 2–3 times per week., via intramuscular or intravenous injection.

Cycle: 4–6 weeks, followed by a 2-week break to check cognitive improvements and response.

Semax + Selank + Dihexa

This combo/stack combines **Semax**, **Selank**, and **Dihexa** to increase focus, reduce anxiety, and improve synaptic connectivity in the brain. **Semax** is known for its cognitive-boosting effects, improving memory and focus, while **Selank** reduces stress and anxiety. **Dihexa** enhances neuroplasticity by promoting new synapse formation, which is beneficial for long-term memory and cognitive resilience. Together, these peptides form a powerful cognitive support stack ideal for professionals, students, or anyone needing sustained mental clarity.

Benefits:

- **Increased Focus and Memory**: Semax improves concentration and mental sharpness, making it easier to stay focused on complex tasks.
- **Stress and Anxiety Reduction**: Selank stabilizes mood, reduces anxiety, and supports a calm, focused state, improving overall cognitive function.
- **Neuroplasticity**: Dihexa supports synapse formation, aiding in memory retention and cognitive flexibility, especially valuable for learning and problem-solving.

Recommended Dosage:

- **Semax :** 300 mcg, 2–3 times daily, via nasal spray or injection. Nasal spray is commonly used for convenience.
- **Selank:** 200–300 mcg, 2–3 times daily, via nasal spray or injection
- **Dihexa**: 10 mg daily, via oral or intramuscular injection.

Cycle: 8–12 weeks, with a 4-week break to allow receptors to reset, especially with Dihexa.

Dihexa + Selank + FGL

This combo/stack combines **Dihexa**, **Selank**, and **FGL**, all of which promote neuroplasticity, cognitive improvement, and memory formation. **Dihexa** is a potent nootropic peptide that enhances synaptic connectivity, while **Selank** reduces anxiety and stress, which often hinder cognitive performance. **FGL** supports neuroplasticity and memory retention, making this stack excellent for long-term cognitive improvement and brain repair.

Benefits:

- **Neuroplasticity Enhancement**: Dihexa and FGL work together to improve synaptic connections, supporting learning and memory.

- **Mood Stabilization**: Selank helps balance mood and reduces stress, allowing for better cognitive function.

- **Memory Support**: This combination aids in forming and retaining new memories, making it ideal for students, professionals, or individuals recovering from brain injuries.

Recommended Dosage:

- **Dihexa:** 10 mg daily, orally or intramuscular injection.

- **Selank:** 200–300 mcg, 2–3 times daily, via nasal spray or injection.

- **FGL:** 100–200 mcg, 1–2 times daily, via subcutaneous injection.

Cycle: 8–12 weeks, with a 4-week break in between cycles to avoid tolerance build-up.

Cerebrolysin + Semax + Epitalon

This combination emphasizes brain repair and neuroprotection, particularly for individuals with neurodegenerative diseases or cognitive decline. **Cerebrolysin** and **Semax** stimulate brain repair and cognitive enhancement, while **Epitalon** regulates circadian rhythms and melatonin production, supporting both brain function and sleep quality, which is essential for cognitive recovery.

Benefits:

- **Cognitive Enhancement and Repair**: Cerebrolysin improves brain function by stimulating the growth and repair of neurons, making it ideal for both cognitive improvement and neurodegenerative conditions.

- **Focus and Mental Clarity**: Semax boosts mental performance by increasing neurotransmitter levels and improving focus.

- **Sleep Support**: Epitalon regulates melatonin production, ensuring better sleep, which is important for brain repair and cognitive health.

Recommended Dosage:

- **Cerebrolysin:** 5–10 ml, 2–3 times per week.

- **Semax:** 300 mcg, 2–3 times daily, via nasal spray or injection.

- **Epitalon (Injection or Oral):** 1–3 mg per day, via **subcutaneous** injection or **orally**, preferably before bedtime.

Cycle: 4–6 weeks with a 2-week break.

Epitalon + Selank + Dihexa

This combo/stack focuses on improving cognitive function while supporting brain longevity and overall mental well-being. **Epitalon** improves sleep quality and regulates circadian rhythms, which are essential for cognitive recovery and neuroprotection. **Selank** reduces anxiety and improves mental clarity, while **Dihexa** aids synaptic connections, promoting long-term brain health and cognitive improvement.

Benefits:

- **Cognitive Longevity:** Epitalon regulates sleep and circadian rhythms, supporting long-term brain health.
- **Reduced Anxiety and Stress:** Selank promotes a calm mental state, improving focus and reducing cognitive stress.
- **Neuroplasticity Support:** Dihexa improves synaptic formation, helping with learning, memory retention, and cognitive flexibility.

Recommended Dosage:

- **Epitalon (Injection or Oral):** 1–3 mg daily, via subcutaneous injection or orally, preferably taken before sleep.
- **Selank:** 200–300 mcg, 2–3 times daily, via nasal spray or injection
- **Dihexa:** 10 mg daily, orally or via intramuscular injection.

Cycle: 8–12 weeks, followed by a 4-week break to assess cognitive improvements.

Semax + CJC-1295 + GHRP-2

This combo/stack focuses on combining cognitive aids with growth hormone support to improve both brain function and physical recovery. **Semax** sharpens mental clarity and memory, while **CJC-1295** and **GHRP-2** stimulate growth hormone release, aiding in overall brain and body recovery. This stack is useful for individuals seeking to improve cognitive performance while benefiting from the regenerative effects of growth hormone.

Benefits:

- **Mental Clarity and Focus:** Semax increases mental sharpness and helps improve memory.
- **Growth Hormone Support:** CJC-1295 and GHRP-2 aid in recovery from brain injuries and cognitive decline by promoting tissue repair and neurogenesis.

- **Overall Cognitive and Physical Recovery**: The growth hormone peptides work synergistically with Semax to improve both brain and body health.

Recommended Dosage:

- **Semax:** 300 mcg, 2–3 times daily, intranasally.
- **CJC-1295:** 1000 mcg twice weekly, via subcutaneous injection.
- **GHRP-2** 100–300 mcg, 1–2 times daily, via subcutaneous injection.

Cycle: 8–12 weeks, followed by a 4-week break to reset growth hormone receptors.

Dihexa + Orexin A + FGL

This stack is an advanced combination for brain health, combining **Dihexa** for synaptic connectivity, **Orexin A** for wakefulness, and **FGL** (a neural cell adhesion molecule mimetic) to enhance learning and memory. **Dihexa** supports neuroplasticity, improving learning capacity and mental clarity. **Orexin A** promotes wakefulness, combating daytime fatigue and brain fog. **FGL** supports memory retention, making this stack particularly valuable for individuals seeking to improve long-term memory and sustained focus throughout the day.

Benefits:

- **Neuroplasticity and Cognitive Flexibility**: Dihexa aids synaptic connections, improving learning speed and mental flexibility.
- **Increased Alertness and Energy**: Orexin A reduces fatigue, promotes sustained energy, and enhances mental endurance, making it easier to stay alert for long periods.
- **Memory Enhancement**: FGL aids in memory consolidation and retention, supporting both short- and long-term memory.

Recommended Dosage:

- **Dihexa:** 10 mg daily, orally or intramuscularly.
- **Orexin A:** 10–20 mg as needed, intranasally, typically taken in the morning.
- **FGL:** 100–200 mcg daily, via subcutaneous or intramuscular injection.

Cycle: 8–12 weeks, with periodic breaks for Orexin A to prevent receptor tolerance and maintain cognitive benefits.

Semax + PE-22-28 + Orexin A

This combo/stack combines **Semax**, **PE-22-28**, and **Orexin A** for cognitive support, memory enhancement, and alertness. **Semax** improves focus and cognitive performance, while **PE-22-28** (an analog of brain-derived neurotrophic factor) promotes brain cell survival and neuroplasticity. **Orexin A** enhances wakefulness and mental energy, making this stack ideal for individuals looking to boost alertness and cognitive clarity throughout the day.

Benefits:

- **Improved Cognitive Performance**: Semax sharpens focus, improves mental clarity, and enhances memory retention, making it easier to tackle complex tasks.
- **Neuroplasticity and Brain Cell Health**: PE-22-28 supports the growth and survival of brain cells, aiding in memory formation and cognitive resilience.
- **Increased Alertness and Wakefulness**: Orexin A naturally promotes wakefulness and sustained energy levels, reducing cognitive fatigue and enhancing mental endurance.

Method of Delivery and Dosage:

- **Semax**: 300 mcg, 2–3 times daily, via Nasal Spray or Injection.
- **PE-22-28:** 100–200 mcg, 1–2 times daily, subcutaneous injection.
- **Orexin A**: 10–20 mg, intranasally, typically administered in the morning or during periods of cognitive fatigue.

Cycle: 8–10 weeks with a break of 4 weeks, especially for Orexin A to avoid receptor tolerance and maintain its effectiveness.

6.4 Peptides Stacks/Combos for Longevity and Anti-Aging

Epitalon + Thymalin + GHK-Cu

This combo/stack focuses on promoting longevity and overall vitality through **Epitalon, Thymalin,** and **GHK-Cu. Epitalon** is known for its ability to activate telomerase, which helps lengthen telomeres and delay cellular aging. **Thymalin** improves immune function and helps reverse some of the age-related immune decline. **GHK-Cu** is a copper peptide that supports cellular regeneration, wound healing, and skin health, making this stack a powerful combination for individuals looking to increase lifespan.

Benefits:

- **Telomere Extension**: Epitalon stimulates telomerase, helping to lengthen telomeres, which are crucial for protecting cells from aging.
- **Immune Support**: Thymalin boosts immune function, which typically declines with age, helping to protect against age-related diseases and infections.
- **Cellular Regeneration and Skin Health**: GHK-Cu improves skin elasticity, reduces wrinkles, and promotes tissue repair, improving both internal and external signs of aging.

Recommended Dosage:

- **Epitalon**: 1–3 mg daily, injected subcutaneously or intramuscularly, for 10–20 days. This cycle can be repeated every 6 months.

- **Thymalin**: 10–20 mg daily for 5–10 days, injected subcutaneously.

- **GHK-Cu**: 2–5 mg daily, injected subcutaneously, or applied **topically** as a cream at a concentration of **0.5–1%**.

Cycle: 10–20 days for Epitalon and Thymalin, with longer continuous use of GHK-Cu (up to 4–6 weeks). Epitalon and Thymalin cycles can be repeated every 6–12 months or once every year.

Epitalon + BPC-157 + TB-500

This combo/stack utilizes **Epitalon** for telomere maintenance and longevity, **BPC-157** for tissue repair and anti-inflammatory effects, and **TB-500** to support neural and connective tissue health. **Epitalon** is known for its role in activating telomerase, which may help delay cellular aging in the brain. **BPC-157** promotes brain resilience and repair by reducing inflammation, and **TB-500** supports neural recovery, particularly for individuals prone to cognitive fatigue or inflammation-related brain fog. Together, these peptides form a potent anti-aging combo/stack that helps protect brain health over the long term.

Benefits:

- **Telomere Maintenance for Longevity**: Epitalon activates telomerase, supporting cellular health and delaying aging at the DNA level, promoting cognitive longevity.

- **Neural Repair and Resilience**: BPC-157 reduces inflammation and improves neural recovery, protecting brain function over time.

- **Support for Connective Tissues and Anti-Inflammation**: TB-500 works synergistically with BPC-157 to promote tissue repair and mitigate inflammation, which is useful for reducing cognitive fatigue.

Recommended Dosage:

- **Epitalon**: 1–3 mg daily for 10–20 days, injected subcutaneously, preferably administered in the evening. This cycle can be repeated every 6 months.

- **BPC-157**: 200–500 mcg daily, injected subcutaneously.

- **TB-500**: 2–5 mg weekly, injected subcutaneously.

- **Cycle:** 8–12 weeks with a break of 4 weeks for BPC-**157** and **TB-500**.

Epitalon + Humanin + GHK-Cu

This longevity combo/stack includes **Epitalon** for telomere health, **Humanin** to combat oxidative stress and protect brain cells, and **GHK-Cu** to support cellular regeneration and collagen production. **Epitalon** helps slow down cellular aging, while **Humanin** acts as a neuroprotective peptide, reducing cellular stress and supporting mitochondrial health. **GHK-Cu** further promotes cell repair and reduces inflammation, making this stack beneficial for individuals seeking to maintain cognitive resilience and brain health as they age.

Benefits:

- **Cellular Longevity and Telomere Support**: Epitalon aids in maintaining telomere length, delaying cellular aging and supporting cognitive health.

- **Mitochondrial Protection and Stress Reduction**: Humanin enhances mitochondrial function, reducing oxidative stress and supporting brain cell survival, critical for longevity.

- **Cellular Regeneration and Inflammation Reduction**: GHK-Cu promotes collagen production and tissue repair, reducing inflammation that can impair brain health.

Recommended Dosage:

- **Epitalon:** 1–3 mg daily for 10-20 days, injected subcutaneously, taken in the evening to align with natural circadian rhythms. This cycle can be repeated once a year.

- **Humanin**: 5 mg daily, injected subcutaneously, to support mitochondrial function.

- **GHK-Cu**: 2–5 mg daily, injected subcutaneously or as a **0.5–1% topical** serum.

Cycle: 8–12 weeks, followed by a 4–6 week break, particularly for Epitalon and Humanin.

MOTS-C + Humanin + SS-31 (Elamipretide)

This combo/stack focuses on mitochondrial health and cellular energy, which helps in slowing down the aging process. **MOTS-C** and **Humanin** are mitochondrial peptides that increases energy production and protect cells from oxidative stress. **SS-31 (Elamipretide)** is a mitochondrial-targeting peptide that helps improve mitochondrial function, reduces inflammation, and protects cells from age-related damage, making this stack useful for improving longevity at the cellular level.

Benefits:

- **Mitochondrial Health and Energy**: MOTS-C and Humanin enhance mitochondrial function, supporting higher energy levels and reducing the risk of age-related fatigue and diseases.

- **Protection from Cellular Damage**: SS-31 protects mitochondria from oxidative stress and reduces inflammation, which are major contributors to aging.

- **Improved Lifespan and Healthspan**: Together, these peptides support longer, healthier lives by addressing mitochondrial dysfunction, one of the hallmarks of aging.

Recommended Dosage:

- **MOTS-C**: 10–15 mg weekly, divided into 2–3 doses, injected subcutaneously.

- **Humanin**: 5 mg daily, injected subcutaneously.

- **SS-31**: 5–10 mg daily, injected subcutaneously.

Cycle: 8–12 weeks of continuous use, followed by a 4-week break.

Epitalon + CJC-1295 + GHRP-2

This combo/stack targets both anti-aging and hormone optimization by combining **Epitalon, CJC-1295,** and **GHRP-2. Epitalon** extends lifespan by activating telomerase and lengthening telomeres, while **CJC-1295** and **GHRP-2** stimulate natural growth hormone production, promoting tissue repair, fat loss, and muscle preservation, all of which are important for healthy aging.

Benefits:

- **Growth Hormone Stimulation**: CJC-1295 and GHRP-2 increase growth hormone levels, which decline with age, helping to improve muscle mass, reduce fat, and support tissue repair.
- **Telomere Protection**: Epitalon helps protect telomeres, delaying cellular aging and promoting longevity.
- **Improved Body Composition**: This stack helps maintain a healthy balance of lean muscle and fat, even as aging slows down metabolism.

Recommended Dosage:

- **Epitalon**: 1–3 mg daily for 10–20 days, injected subcutaneously.
- **CJC-1295**: 1000 mcg twice weekly, subcutaneously.
- **GHRP-2**: 100–200 mcg, 1–2 times daily, injected subcutaneously.

Cycle: 10–12 weeks with a break of 4–6 weeks. Epitalon is cycled every 6 months, while CJC-1295 and GHRP-2 can be used for longer periods, with periodic breaks.

GHK-Cu + BPC-157 + TB-500

This combo/stack is focused on tissue repair, wound healing, and overall cellular health. **GHK-Cu** promotes collagen production and skin regeneration, **BPC-157** accelerates tissue repair and reduces inflammation, and **TB-500** supports recovery from injuries and promotes muscle and tendon healing. Together, they create a useful anti-aging and recovery combo/stack, helping the body maintain youthful tissue and repair age-related damage.

Benefits:

- **Skin and Tissue Repair**: GHK-Cu improves skin elasticity and reduces wrinkles, while BPC-157 and TB-500 help heal injuries and reduce inflammation.
- **Accelerated Healing**: BPC-157 and TB-500 work synergistically to speed up recovery from injuries and surgeries, supporting long-term tissue health.

- **Anti-Aging and Longevity**: GHK-Cu and BPC-157 have regenerative properties that promote overall tissue health, improving both internal and external signs of aging.

Recommended Dosage:

- **GHK-Cu**: 2–5 mg daily, injected subcutaneously, or applied **topically** as a **0.5–1%** cream.
- **BPC-157**: 200–500 mcg daily, subcutaneously.
- **TB-500**: 2–5 mg weekly, subcutaneously.

Cycle: 8–12 weeks for all three peptides, with periodic breaks.

Thymalin + Epitalon + GHRP-6

This longevity combo/stack combines the immune-boosting and anti-aging benefits of **Thymalin** and **Epitalon** with the growth hormone-stimulating effects of **GHRP-6**. **Thymalin** increases immune function and reduces inflammation, while **Epitalon** promotes healthy aging by protecting telomeres. **GHRP-6** increases natural growth hormone levels, supporting fat loss, muscle retention, and overall vitality as you age.

Benefits:

- **Telomere Maintenance and Longevity**: Epitalon helps preserve telomeres, promoting cellular longevity and protecting against age-related decline.
- **Immune System Boost**: Thymalin strengthens the immune system, helping the body fight off infections and age-related diseases.
- **Growth Hormone Release**: GHRP-6 stimulates GH production, improving body composition and supporting healthy aging.

Recommended Dosage:

- **Thymalin**: 10–20 mg daily for 5–10 days, injected subcutaneously.
- **Epitalon**: 1–3 mg daily for 10–20 days, injected subcutaneously.
- **GHRP-6**: 100–300 mcg daily, injected subcutaneously.

Cycle: 10–20 days for Epitalon and Thymalin, repeated every 6 months. GHRP-6 can be used for longer cycles (8–12 weeks), followed by a break.

6.5 Peptides Stacks/Combos for Sexual Health

PT-141 + Kisspeptin + Melanotan II

This combo/stack combines **PT-141**, **Kisspeptin**, and **Melanotan II** to boost sexual arousal and improve sexual function in both men and women. **PT-141** is a known libido-increasing peptide that acts on the melanocortin receptors in the brain, improving sexual desire and function. **Kisspeptin** supports fertility by stimulating gonadotropin-releasing hormone (GnRH), which in turn triggers the production of luteinizing hormone (LH) and follicle-stimulating hormone (FSH), enhancing reproductive health. **Melanotan II** offers additional libido enhancement and helps regulate sexual response.

Benefits:

- **Increased Libido**: PT-141 and Melanotan II both stimulate sexual desire and arousal, improving overall sexual experience.

- **Sexual Function**: PT-141 improves erectile function in men and arousal in women, making it effective for treating sexual dysfunction.

- **Fertility Support**: Kisspeptin aids in reproductive hormone regulation, improving fertility in both men and women.

Recommended Dosage:

- **PT-141**: 1–2 mg per injection, taken 30–60 minutes before sexual activity, injected subcutaneously.

- **Kisspeptin**: 100–200 mcg daily, injected subcutaneously, to support fertility.

- **Melanotan II**: 0.25–1 mg per injection, taken 1–2 times per week, injected subcutaneously.

Cycle: Used on-demand for PT-141 and Melanotan II. Kisspeptin is typically used in **4–6-week cycles** for fertility.

PT-141 + CJC-1295 + Ipamorelin

This combo/stack is designed for individuals looking to improve their sexual health and overall hormonal balance. **PT-141** focuses on improving libido and sexual function, while **CJC-1295** and **Ipamorelin** work together to increase growth hormone levels, which can improve energy, vitality, and sexual performance. This combination is beneficial for men and women seeking to improve their sexual wellness alongside overall health and vitality.

Benefits:

- **Increased Sexual Desire and Performance**: PT-141 improves libido and improves sexual function in both men and women.

- **Improved Vitality and Hormonal Balance**: CJC-1295 and Ipamorelin increase growth hormone levels, supporting better energy, mood, and sexual performance.

- **Better Recovery**: Increased growth hormone levels improve recovery and overall physical and mental health, which can also support sexual health.

Recommended Dosage:

- **PT-141:** 1–2 mg per injection, taken 30–60 minutes before sexual activity, injected subcutaneously.

- **CJC-1295**: 1000 mcg twice weekly, injected subcutaneously.
- **Ipamorelin**: 200–300 mcg, 1–2 times daily, injected subcutaneously.

Cycle: 8–12 weeks for CJC-1295 and Ipamorelin, with breaks. PT-141 is used on-demand.

Gonadorelin + PT-141 + MK-677

This combo/stack combines **Gonadorelin**, **PT-141**, and **MK-677** to optimize sexual health and hormonal balance in **men**. **Gonadorelin** stimulates the production of LH and FSH, leading to an increase in natural testosterone levels, improving libido and sexual performance. **PT-141** increases sexual desire, and **MK-677** boosts growth hormone levels, which support muscle mass, energy, and overall sexual health.

Benefits:

- **Testosterone Boost**: Gonadorelin increases natural testosterone production, improving sexual performance and energy in men.
- **Improved Libido and Arousal**: PT-141 directly stimulates the brain's melanocortin receptors, increasing sexual desire and function.
- **Improved Recovery and Body Composition**: MK-677 increases growth hormone levels, supporting better recovery, fat loss, and overall vitality.

Recommended Dosage:

- **Gonadorelin**: 100–200 mcg daily, injected subcutaneously or intramuscularly.
- **PT-141:** 1–2 mg per injection, taken 30–60 minutes before sexual activity, injected subcutaneously.
- **MK-677:** 10–25 mg daily, taken orally.

Cycle: 8–12 weeks, with a break of 4–6 weeks for Gonadorelin and MK-677. PT-141 is used as needed.

Kisspeptin + CJC-1295 + Ipamorelin

This combo/stack focuses on optimizing reproductive health and sexual function by using **Kisspeptin** to stimulate reproductive hormones, while **CJC-1295** and **Ipamorelin** increase growth hormone levels, supporting overall vitality. This combination is particularly effective for **women** looking to improve libido, fertility, and hormonal balance, especially during menopause or periods of hormonal imbalance.

Benefits:

- **Fertility:** Kisspeptin supports ovulation and hormone balance, improving fertility in women.
- **Improved Sexual Health and Libido**: Kisspeptin increases sexual desire, while CJC-1295 and Ipamorelin boost energy and mood, indirectly supporting sexual health.
- **Better Hormonal Balance**: This stack regulates reproductive hormones and supports overall well-being, particularly in women undergoing menopause or experiencing hormonal imbalances.

Recommended Dosage:

- **Kisspeptin**: 100–200 mcg daily, injected subcutaneously.
- **CJC-1295**: 1000 mcg twice weekly, injected subcutaneously.
- **Ipamorelin**: 200–300 mcg, 1–2 times daily, injected subcutaneously.

Cycle: 8–12 weeks with periodic breaks for hormonal regulation.

PT-141 + Melanotan II + CJC-1295

This combo/stack is ideal for individuals who want to improve both sexual health and body composition. **PT-141** increases libido and sexual performance, **Melanotan II** provides additional libido support and improves skin pigmentation, while **CJC-1295** increases growth hormone levels, promoting better recovery and overall vitality.

Benefits:

- **Sexual Desire and Performance**: PT-141 and Melanotan II both work on the melanocortin receptors, significantly boosting libido and sexual satisfaction.
- **Improved Body Composition**: CJC-1295 stimulates growth hormone release, which aids in fat loss and muscle preservation.
- **Skin Pigmentation**: Melanotan II helps users achieve a tan while improving sexual health.

Recommended Dosage:

- **PT-141**: 1–2 mg per injection, taken 30–60 minutes before sexual activity, injected subcutaneously.
- **Melanotan II**: 0.25–1 mg per injection, 1–2 times per week, injected subcutaneously.
- **CJC-1295**: 1000 mcg twice weekly, injected subcutaneously.

Cycle: 12 weeks with periodic breaks for CJC-1295 and Melanotan II. PT-141 can be used as needed.

6.6 Peptides Stacks/Combos for Immunity

Thymosin Alpha-1 + LL-37 + VIP

This combo/stack is powerful for boosting the immune system and fighting infections. **Thymosin Alpha-1** stimulates T-cell production, improving immune response. **LL-37** is an antimicrobial peptide that kills bacteria and viruses, while **VIP** (Vasoactive Intestinal Peptide) reduces inflammation and improves lung health, making this combination particularly useful during flu seasons or for individuals with chronic immune challenges.

Benefits:

- **Immune Boosting**: Thymosin Alpha-1 strengthens the immune system by increasing T-cell activity.

- **Antimicrobial Action**: LL-37 directly combats bacteria, viruses, and fungi, making it useful for both prevention and treatment of infections.
- **Lung and Respiratory Health**: VIP reduces inflammation in the lungs and supports healthy respiratory function.

Recommended Dosage:

- **Thymosin Alpha-1:** 1.6–3.2 mg weekly, injected subcutaneously.
- **LL-37:** 100–300 mcg daily, injected subcutaneously.
- **VIP**: 50 mcg sprayed within each nostril up to 4 times per day.

Cycle: 4–6 weeks during times of immune suppression or increased infection risk.

Thymosin Alpha-1 + BPC-157 + SS-31

This combo/stack is designed to improve immunity and promote healing. **Thymosin Alpha-1** boosts immune function, **BPC-157** promotes tissue repair and reduces inflammation, and **SS-31** supports mitochondrial health, reducing oxidative stress and protecting the immune system from damage.

Benefits:

- **Immune Function Support**: Thymosin Alpha-1 improves immune response, helping fight infections and boosting general immunity.
- **Tissue Healing and Repair**: BPC-157 helps heal tissues, especially useful for those recovering from surgery or injury.
- **Mitochondrial Protection**: SS-31 reduces oxidative damage, supporting both immune health and overall vitality.

Recommended Dosage:

- **Thymosin Alpha-1**: 1.6–3.2 mg weekly, injected subcutaneously.
- **BPC-157**: 200–500 mcg daily, injected subcutaneously.
- **SS-31:** 5–10 mg daily, injected subcutaneously.

Cycle: 8–12 weeks with periodic breaks to monitor immune function.

VIP + LL-37 + SS-31

This immunity combo/stack combines **VIP (Vasoactive Intestinal Peptide)**, **LL-37**, and **SS-31 (Elamipretide)** to support immune resilience, reduce inflammation, and protect mitochondrial health. **VIP** acts as a powerful anti-inflammatory agent, improving lung and respiratory health, while **LL-37** provides antimicrobial action against pathogens. **SS-31 (Elamipretide)** supports mitochondrial function, which is crucial for immune cell energy and resilience, particularly in the face of chronic infections or inflammatory conditions.

Benefits:

- **Anti-Inflammatory and Respiratory Support**: VIP reduces inflammation in respiratory tissues, making it beneficial for individuals with chronic respiratory issues or those exposed to pathogens.

- **Antimicrobial Defense**: LL-37 offers broad-spectrum antimicrobial effects, protecting against bacterial, viral, and fungal infections.

- **Mitochondrial Protection and Immune Resilience**: SS-31 supports mitochondrial health, ensuring immune cells have the energy needed to respond effectively to infections and inflammation.

Method of Delivery and Dosage:

- **VIP**: 100–500 mcg daily, injected subcutaneously or intranasally (50 mcg sprayed within each nostril up to 4 times per day)

- **LL-37**: 100–300 mcg daily, injected subcutaneously.

- **SS-31**: 5–10 mg daily, injected subcutaneously.

Cycle: 8–12 weeks, with a 4-week or more.

Thymosin Alpha-1 + KPV + ARA-290

This immune combo/stack utilizes **Thymosin Alpha-1**, **KPV**, and **ARA-290** to strengthen the immune system, reduce inflammation, and alleviate pain associated with chronic inflammation. **Thymosin Alpha-1** aids T-cell activity and immune response, **KPV** reduces inflammatory responses, particularly in the gut, and **ARA-290** provides pain relief and supports nerve health by reducing inflammation in peripheral tissues. This combination is beneficial for those looking to support immune health and mitigate symptoms of autoimmune or inflammatory conditions.

Benefits:

- **Immune Function**: Thymosin Alpha-1 boosts the body's immune defenses by increasing T-cell production and response to infections.

- **Inflammation and Pain Reduction**: KPV has strong anti-inflammatory effects, particularly beneficial for gut health, while ARA-290 provides relief from inflammatory pain and promotes tissue healing.

- **Improved Recovery from Autoimmune and Chronic Conditions**: This combination supports immune balance, making it effective for managing symptoms of autoimmune diseases and chronic inflammation.

Recommended Dosage:

- **Thymosin Alpha-1**: 1.6–3.2 mg weekly, injected subcutaneously.
- **KPV**: 200–400 mcg daily, injected subcutaneously.
- **ARA-290:** 4 mg, 2–3 times weekly, injected subcutaneously.

Cycle: 8–12 weeks, with periodic breaks to evaluate immune response, especially for Thymosin Alpha-1.

Thymosin Alpha-1 + LL-37 + BPC-157

This immune and recovery combo/stack combines **Thymosin Alpha-1**, **LL-37**, and **BPC-157** to bolster the immune system, fight infections, and promote healing of damaged tissues. **Thymosin Alpha-1** supports immune regulation, **LL-37** provides antimicrobial protection against pathogens, and **BPC-157** aids tissue repair and reduces inflammation. This stack is useful for individuals recovering from illness, injury, or surgery who need strong immune and tissue support.

Benefits:

- **Immune Response**: Thymosin Alpha-1 strengthens immune defenses, increasing resistance to infections.
- **Antimicrobial and Infection Control**: LL-37 fights a range of pathogens, including bacteria and viruses, reducing the likelihood of infections.
- **Accelerated Healing and Reduced Inflammation**: BPC-157 supports tissue repair and reduces inflammation, aiding recovery from injuries or surgical procedures.

Recommended Dosage:

- **Thymosin Alpha-1**: 1.6–3.2 mg weekly, injected subcutaneously.
- **LL-37**: 100–300 mcg daily, injected subcutaneously.
- **BPC-157**: 200–500 mcg daily, injected subcutaneously.

Cycle: 8–12 weeks, with a 4-week break to assess immune function and response.

6.7 Peptides Stacks/Combos for Skin, Hair and Aesthetics

GHK-Cu + BPC-157 + Epitalon

This combo/stack is designed to improve skin health, reduce wrinkles, and promote collagen production. **GHK-Cu** is known for its powerful anti-aging and skin-repairing properties, **BPC-157** accelerates tissue repair and wound healing, and **Epitalon** supports overall skin regeneration by improving melatonin regulation and boosting telomerase activity, which helps to reduce cellular aging.

Benefits:

- **Increased Collagen Production**: GHK-Cu stimulates collagen synthesis, helping to reduce wrinkles and improve skin elasticity.

- **Tissue Repair and Healing**: BPC-157 promotes skin regeneration and reduces inflammation, improving overall skin health.

- **Anti-Aging and Longevity**: Epitalon supports cellular repair and helps regulate sleep patterns, indirectly improving skin health.

Recommended Dosage:

- **GHK-Cu**: 2–5 mg daily as a topical serum (0.5–1% concentration).

- **BPC-157**: 200–500 mcg daily, injected subcutaneously.

- **Epitalon**: 1–3 mg daily for **10–20 days**, injected subcutaneously once yearly. This cycle can be repeated every **6–12 months** for long-term sleep.

Cycle: 8–12 weeks for **GHK-Cu & BPC-157**, with a break of 4 weeks between cycles.

GHK-Cu + PTD-DBM + Argireline

This cosmetic combo/stack combines **GHK-Cu**, **PTD-DBM**, and **Argireline** to benefit the skin and hair aesthetics. **GHK-Cu** is renowned for its skin-rejuvenating properties, promoting collagen synthesis, improving skin elasticity, and aiding in wound healing. **PTD-DBM** targets hair health, supporting follicle regeneration and encouraging hair growth, making it effective for addressing thinning hair.

Argireline serves as a non-invasive anti-wrinkle solution, relaxing facial muscles and smoothing fine lines without the need for injections. Together, this stack enhances skin quality, supports hair growth, and offers anti-aging benefits, making it a versatile solution for overall cosmetic improvement.

Benefits:

- **Enhanced Skin Texture and Elasticity**: GHK-Cu stimulates collagen production, which smooths fine lines and firms the skin, improving overall texture.

- **Wrinkle Reduction**: Argireline relaxes facial muscles, reducing the depth of wrinkles and creating a smoother appearance, especially around expression-prone areas.

- **Promotes Hair Growth and Scalp Health**: PTD-DBM supports hair follicle activity, encouraging hair growth in thinning areas and improving scalp condition.

Recommended Dosage:

- **GHK-Cu**: 2–5 mg daily topically at a concentration of 0.5–1% in a serum for skin application.

- **PTD-DBM**: Applied topically to the scalp at a concentration of 0.1–0.5% for hair growth support.

- **Argireline:** Applied topically daily to targeted areas in concentrations of 5–10% as a cream or serum.

Cycle: GHK-Cu and Argireline can be used continuously as part of a daily skincare routine. For **PTD-DBM**, an 8–12-week cycle is ideal, followed by a break of 4 weeks before resuming to assess hair growth and follicle health.

GHK-Cu + CJC-1295 + Ipamorelin

This combo/stack combines **GHK-Cu** for its anti-aging and skin-regeneration properties, with **CJC-1295** and **Ipamorelin** to promote growth hormone release, enhancing skin elasticity, muscle tone, and fat reduction. Together, these peptides promote both internal and external rejuvenation.

Benefits:

- **Improved Skin Elasticity and Texture**: GHK-Cu increases collagen production, making skin firmer and reducing wrinkles.
- **Growth Hormone Support**: CJC-1295 and Ipamorelin increase growth hormone levels, helping with fat loss, muscle retention, and overall vitality.
- **Youthful Appearance**: This combination improves overall skin health and supports a more youthful appearance.

Method of Delivery and Dosage:

- **GHK-Cu**: 2–5 mg daily as a topical serum (0.5–1% concentration).
- **CJC-1295**: 1000 mcg twice weekly, injected subcutaneously.
- **Ipamorelin**: 200–300 mcg, 1–2 times daily, injected subcutaneously.

Cycle: 8–12 weeks for CJC-1295 and Ipamorelin. GHK-Cu can be used continuously for longer periods.

BPC-157 + GHRP-2 + GHK-Cu

This combo/stack is ideal for skin repair, tissue healing, and overall anti-aging effects. **BPC-157** promotes rapid healing of skin, muscle, and connective tissue, **GHRP-2** stimulates growth hormone release to support skin elasticity and muscle tone, and **GHK-Cu** provides powerful anti-aging effects by promoting collagen production and skin regeneration.

Benefits:

- **Tissue and Skin Repair**: BPC-157 accelerates healing and reduces inflammation, making it ideal for individuals recovering from injuries or surgeries.
- **Growth Hormone Release**: GHRP-2 boosts growth hormone, improving muscle tone and skin elasticity.
- **Anti-Aging**: GHK-Cu improves skin texture and appearance by stimulating collagen production.

Method of Delivery and Dosage:

- **BPC-157**: 200–500 mcg daily, injected subcutaneously.
- **GHRP-2**: 100–300 mcg daily, injected subcutaneously.
- **GHK-Cu**: 2–5 mg daily as a topical serum (0.5–1% concentration).

Cycle: 8–12 weeks with breaks for GHRP-2 and GHK-Cu.

6.8 Key Considerations for Peptide Combos/Stacking

- Choose peptides that complement each other in terms of how they work. For example, stacking/combining peptides that both promote growth hormone release with peptides that improve tissue repair can lead to better muscle recovery and growth.
- Peptide stacks should be cycled to prevent the body from developing a tolerance or diminishing returns. A typical cycle might last 4–8 weeks, followed by a break of a few weeks before starting again. This ensures that the peptides remain effective and reduces the risk of side effects from prolonged use.
- When stacking peptides, it's important to adjust the dosages to ensure you are not overloading your system. The recommended dosages for each peptide in a stack may be lower than if you were taking them individually, as the combined effect of the stack is more powerful.
- Keep track of how your body responds to the peptide stack, especially if you are new to peptide therapy.

CHAPTER 7. PEPTIDES AND LIFESTYLE

Peptides work best when integrated into a healthy lifestyle. To maximize the benefits of peptide therapy, it's important to support your body with the right nutrition, exercise, recovery strategies and manage your expectations properly.

7.1 Nutrition, Exercise and Recovery

7.1.1 Nutrition

Protein Intake

Many peptides, particularly those used for muscle growth and recovery (such as CJC-1295, Ipamorelin, or IGF-1 LR3), rely on adequate protein intake to support muscle protein synthesis. Aim to consume 1.0–1.2 grams of protein per pound of body weight daily. This can come from sources like lean meats, fish, eggs, dairy, or plant-based protein powders.

Healthy Fats

Hormonal peptides that influence testosterone, estrogen, or growth hormone levels will work better if your body has access to healthy fats. Omega-3 fatty acids (from fish, flaxseeds, or walnuts) support hormone production, reduce inflammation, and improve overall cellular health.

Antioxidants

Peptides like GHK-Cu and BPC-157 promote tissue repair and reduce inflammation. To support this process, focus on eating a diet rich in antioxidants such as fruits, vegetables, nuts, and seeds help combat oxidative stress, which can impair recovery and cellular health.

Hydration

Staying hydrated is essential for muscle recovery, tissue repair, and overall health. Drink at least 8–10 glasses of water per day and consider increasing this amount if you're using peptides for performance or fat loss, as they help to improve metabolic activity.

7.1.2 Exercise

Strength Training

For muscle growth peptides, engaging in regular resistance training is important. Focus on compound movements (such as squats, deadlifts, and presses) that work large muscle groups. Aim for 3–5 sessions per week, with progressive overload to continuously challenge your muscles.

Cardiovascular Exercise

For people using fat-loss peptides like AOD-9604, Semaglutide, etc. incorporating cardio is important. High-intensity interval training (HIIT) is particularly effective for maximizing fat loss, while steady-state cardio can support overall cardiovascular health and endurance.

Recovery Sessions

Peptides like BPC-157 and TB-500 improve recovery. Complement this by incorporating low-intensity recovery activities (such as yoga, swimming, or walking) to promote circulation, reduce inflammation, and improve muscle repair.

7.1.3 Recovery

Sleep

Peptides like DSIP, Epitalon, or CJC-1295 optimize recovery during sleep. Aim for 7–9 hours of quality sleep each night. Sleep is when your body repairs muscles, processes information, and balances hormone levels. Skimping on sleep can hinder your progress, no matter how well your peptides are working.

Stress Management

Peptides like Selank or Semax can help manage stress but integrating other stress-reducing practices (such as meditation, deep breathing, or mindfulness) into your routine can further support peptide efficacy. High stress levels can disrupt hormonal balance, impair cognitive function, and lead to inflammation, all of which counteract the benefits of peptide therapy.

7.2 Managing Your Expectations

Understanding the difference between short-term and long-term benefits is important when using peptides, as different peptides offer results on different timelines.

7.2.1 Short-Term Benefits (Within Days to Weeks)

Energy and Focus:

Peptides like **Semax** or **Selank** often deliver noticeable improvements in focus, cognitive function, and mood within a few days. Individuals are likely to experience improved clarity, reduced anxiety, and better mental performance relatively quickly.

Sleep Improvements

Peptides like **DSIP** and **Epitalon** can improve sleep quality within the first week of use. Users often report falling asleep faster, experiencing fewer awakenings, and waking up more refreshed within the first few nights.

Appetite Suppression

For fat-loss peptides like **Semaglutide** or **Tirzepatide**, appetite suppression can occur within the first few doses, making it easier to reduce calorie intake and start losing weight.

7.2.2 Long-Term Benefits (Within Months)

Muscle Growth and Fat Loss

Peptides like **CJC-1295**, **Ipamorelin**, or **IGF-1 LR3** may take 8–12 weeks before significant muscle gains or fat loss are noticeable. Building muscle and burning fat require consistent use combined with proper nutrition and exercise.

Anti-Aging and Skin Health

Peptides like **GHK-Cu** or **Epitalon** support skin rejuvenation and anti-aging effects, but these changes occur over several months. You might notice subtle improvements in skin texture, elasticity, and wrinkles, but dramatic changes take time.

Longevity and Immune Support

Peptides like **Thymosin Alpha-1** and **Epitalon** that support immune function or cellular longevity often provide benefits over the long term. Better immune defense or improvements in age-related symptoms might not be immediately noticeable but contribute to better long-term health.

7.2.3 Balancing Expectations

Achieving long-term results with peptides requires consistent use over an extended period. Stick to the recommended cycles and dosages, even if you don't see immediate changes.

Peptides are not magic solutions. Their effects are amplified when combined with healthy lifestyle practices, including balanced nutrition, regular exercise, and adequate sleep.

Monitor small improvements over time, whether it's better recovery, slight reductions in body fat, or smoother skin. These incremental changes accumulate into significant results after several months.

CHAPTER 8. CONCLUSION

Peptides have become one of the most exciting advancements in modern medicine, offering a vast range of therapeutic applications, from anti-aging and skin care to fat loss, muscle growth, immune support, cognitive enhancement and brain function, sexual health, and more.

Their ability to target the specific root causes of many health challenges with minimal side effects has made peptide therapy a preferred choice for many individuals, athletes, and medical professionals. As research continues to advance, the potential for peptides in preventive medicine, treatment of chronic conditions, and personalized health solutions will only expand.

This book has covered a wide range of peptides and how they can be stacked/combined for specific results, along with practical guidance for safe preparation and use. Just like any wellness journey, the key to success lies in combining peptide therapy with a healthy lifestyle and understanding how your body responds.

Remember that peptides are powerful, so it's always best to approach them with care. Work with a healthcare professional to help monitor your progress and adjust dosages as needed.

Thank you for reading, and Goodluck!

8.1 Resources for Further Learning and Research

As the field of peptide therapy continues to grow, staying informed about the latest developments, research, and products is important for anyone interested in using peptides. Here are some key resources for further learning and research:

1. Medical Journals and Research Publications

- **PubMed**: This is one of the largest databases of scientific research papers, including many studies on peptide therapy. You can search for specific peptides and review the latest clinical trials and peer-reviewed research.
- **ResearchGate**: A platform where researchers share their publications and findings. It's a great resource for accessing studies on emerging peptide therapies and discussing findings with other professionals in the field.

2. Professional Organizations

- **International Peptide Society (IPS)**: A professional organization dedicated to advancing the field of peptide therapy. They offer educational resources, webinars, and training courses for both healthcare providers and individuals interested in peptide use.
- **American Academy of Anti-Aging Medicine (A4M)**: A global organization that focuses on advancements in anti-aging medicine, including peptide therapies. They host conferences, publish research, and offer certifications in peptide therapy.

3. Educational Websites and Forums

- **Peptide Sciences Blog**: A reputable source for news and updates on peptide research, applications, and safety information.

- **Peptides.org**: Provides detailed explanations of how different peptides work, their benefits, and how they can be integrated into health routines.

- **Fitness and Wellness Forums**: Online communities, such as Reddit's **r/Peptides** or **r/Nootropics**, are excellent places to engage in discussions with other users about their experiences with peptide therapy. These forums often provide practical insights, product reviews, and advice on stacks and combinations.

4. Healthcare Providers and Peptide Specialists

Working with a healthcare provider experienced in peptide therapy is essential for ensuring safe and effective use. Many functional medicine doctors, endocrinologists, and anti-aging specialists are knowledgeable about peptide therapy and can guide you in creating personalized treatment plans.

References

Almeida, J. R. (2024). The Century-Long Journey of Peptide-Based Drugs. *Antibiotics*, *13*(3), 196. https://doi.org/10.3390/antibiotics13030196

Doti, N., & Ruvo, M. (2024). Synthetic Peptides and Peptidomimetics: From Basic Science to Biomedical Applications—Second Edition. *International Journal of Molecular Sciences*, *25*(2), 1083–1083. https://doi.org/10.3390/ijms25021083

Fetse, J., Kandel, S., Mamani, U.-F., & Cheng, K. (2023). *Recent advances in the development of therapeutic peptides*. *44*(7), 425–441. https://doi.org/10.1016/j.tips.2023.04.003

Li, L., Gregory Joseph Duns, Wubliker Dessie, Cao, Z., Ji, X., & Luo, X. (2023). Recent advances in peptide-based therapeutic strategies for breast cancer treatment. *Frontiers in Pharmacology*, *14*. https://doi.org/10.3389/fphar.2023.1052301

Marcin, A. (2023, November 13). *Weight Loss Peptides: Everything You Need to Know*. Healthline; Healthline Media. https://www.healthline.com/health/weight-loss/using-peptides-for-weight

Martini, S., & Davide Tagliazucchi. (2023). *Bioactive Peptides in Human Health and Disease*. *24*(6), 5837–5837. https://doi.org/10.3390/ijms24065837

Naeem, M., Muhammad Inamullah Malik, Umar, T., Ashraf, S., & Ahmad, A. (2022). A Comprehensive Review About Bioactive Peptides: Sources to Future Perspective. *International Journal of Peptide Research and Therapeutics*, *28*(6). https://doi.org/10.1007/s10989-022-10465-3

Ngoc, L. T. N., Moon, J.-Y., & Lee, Y.-C. (2023). Insights into Bioactive Peptides in Cosmetics. *Cosmetics*, *10*(4), 111. https://doi.org/10.3390/cosmetics10040111

Nhàn, T., Yamada, T., & Yamada, K. H. (2023). Peptide-Based Agents for Cancer Treatment: Current Applications and Future Directions. *International Journal of Molecular Sciences*, *24*(16), 12931–12931. https://doi.org/10.3390/ijms241612931

Othman Al Musaimi. (2024). Peptide Therapeutics: Unveiling the Potential against Cancer—A Journey through 1989. *Cancers*, *16*(5), 1032–1032. https://doi.org/10.3390/cancers16051032

Pereira, A. J., Luana, Xing, H., & Conda-Sheridan, M. (2024). Peptide-based therapeutics: challenges and solutions. *Medicinal Chemistry Research*. https://doi.org/10.1007/s00044-024-03269-1

Petre MS, RD (NL), A. (2020, December 3). *Peptides for Bodybuilding: Do They Work, and Are They Safe?* Healthline. https://www.healthline.com/nutrition/peptides-for-bodybuilding

Purohit, K., Reddy, N., & Anwar Sunna. (2024). Exploring the Potential of Bioactive Peptides: From Natural Sources to Therapeutics. *International Journal of Molecular Sciences*, *25*(3), 1391–1391. https://doi.org/10.3390/ijms25031391

Richard, O.-A. (2019). *Bioactive Peptides*. Google Books. https://books.google.com.ng/books?id=JJ_MBQAAQBAJ&lpg=PP1&ots=DzI9Z5uKH5&dq=Bioactive%20peptides%20and%20health.%20(n.d.).%20Frontiers%20in%20Nutrition&lr&pg=PR6#v=onepage&q&f=false

Rivero-Pino, F. (2023). Bioactive food-derived peptides for functional nutrition: Effect of fortification, processing and storage on peptide stability and bioactivity within food matrices. *Food Chemistry, 406*, 135046. https://doi.org/10.1016/j.foodchem.2022.135046

Rossino, G., Marchese, E., Galli, G., Verde, F., Finizio, M., Serra, M., Linciano, P., & Collina, S. (2023). Peptides as Therapeutic Agents: Challenges and Opportunities in the Green Transition Era. *Molecules, 28*(20), 7165. https://doi.org/10.3390/molecules28207165

Sreenivas, S. (2021, March 25). *What Are Peptides?* WebMD. https://www.webmd.com/a-to-z-guides/what-are-peptides

Wang, L., Wang, N., Zhang, W., Cheng, X., Yan, Z., Shao, G., Wang, X., Wang, R., & Fu, C. (2022). Therapeutic peptides: current applications and future directions. *Signal Transduction and Targeted Therapy, 7*(1), 48. https://doi.org/10.1038/s41392-022-00904-4

www.ingramcontent.com/pod-product-compliance
Lightning Source LLC
Chambersburg PA
CBHW082251220526
45469CB00009B/2954